Springer Theses

Recognizing Outstanding Ph.D. Research

Aims and Scope

The series "Springer Theses" brings together a selection of the very best Ph.D. theses from around the world and across the physical sciences. Nominated and endorsed by two recognized specialists, each published volume has been selected for its scientific excellence and the high impact of its contents for the pertinent field of research. For greater accessibility to non-specialists, the published versions include an extended introduction, as well as a foreword by the student's supervisor explaining the special relevance of the work for the field. As a whole, the series will provide a valuable resource both for newcomers to the research fields described, and for other scientists seeking detailed background information on special questions. Finally, it provides an accredited documentation of the valuable contributions made by today's younger generation of scientists.

Theses are accepted into the series by invited nomination only and must fulfill all of the following criteria

- They must be written in good English.
- The topic should fall within the confines of Chemistry, Physics, Earth Sciences, Engineering and related interdisciplinary fields such as Materials, Nanoscience, Chemical Engineering, Complex Systems and Biophysics.
- The work reported in the thesis must represent a significant scientific advance.
- If the thesis includes previously published material, permission to reproduce this must be gained from the respective copyright holder.
- They must have been examined and passed during the 12 months prior to nomination.
- Each thesis should include a foreword by the supervisor outlining the significance of its content.
- The theses should have a clearly defined structure including an introduction accessible to scientists not expert in that particular field.

More information about this series at http://www.springer.com/series/8790

David Chinarro

System Engineering Applied to Fuenmayor Karst Aquifer (San Julián de Banzo, Huesca) and Collins Glacier (King George Island, Antarctica)

Doctoral Thesis accepted by the University of Zaragoza, Spain

Springer

Author
Dr. David Chinarro
Polytechnic School
San Jorge University
Zaragoza
Spain

Supervisors
Prof. José L. Villarroel
Department of Computer and Systems Engineering
University of Zaragoza, Group of Technologies in Harsh Environments (GTE)
Zaragoza
Spain

Prof. José A. Cuchí
Department of Agricultural Sciences and Natural Habitats, Polytechnic School of Huesca
University of Zaragoza
Huesca
Spain

ISSN 2190-5053 ISSN 2190-5061 (electronic)
ISBN 978-3-319-08857-0 ISBN 978-3-319-08858-7 (eBook)
DOI 10.1007/978-3-319-08858-7

Library of Congress Control Number: 2014943252

Springer Cham Heidelberg New York Dordrecht London

Printed on acid-free paper

Springer is part of Springer Science+Business Media (www.springer.com)

Parts of this thesis have been published in the following journal articles:

- **Application of Wavelet Correlation Analysis to the Karst Spring of Fuenmayor. San Julian de Banzo, Huesca, Spain (2011).**
  Authors: David Chinarro, José L. Villarroel and José A. Cuchí.
  Chapter in book: *Advances in Research in Karst Media. Series: Environmental Earth Sciences.*
  Authors: Carrasco, Francisco; Durán Valsero, Juan José; LaMoreaux, James W. (Eds.).
  Publisher: Springer. 1st Edition, 2011, ISBN 978-3-642-12485-3.
- **Wavelet analysis of Fuenmayor karst spring, San Julián de Banzo, Huesca, Spain (2011 and 2012).**
  Authors: D. Chinarro, J. L. Villarroel and J. A. Cuchí.
  Journal: Environmental Earth Sciences ISSN: 1866-6299, doi: 10.1007/s12665-011-1351-y, September 17, 2011. 13 pages.
  Reprinted: April 2012, ISSN 1866-6280, 2012, Volume 65, Number 8, pages 2231–2243.
- **Linear System Techniques Applied to the Fuenmayor Karst Spring, Huesca (Spain) (2013).**
  Authors: J. A. Cuchí, D. Chinarro and J. L. Villarroel.
  Journal : Environmental Earth Sciences.
- **Application of wavelet correlation analysis to the karst spring of Fuenmayor. San Julián de Banzo, Huesca, Spain (2010).**
  Authors: D. Chinarro, J. A. Cuchi and J. L. Villarroel.
  Congress: 4th International Symposium on Karst (ISKA-2010). Málaga, Spain, April 2010.
- **Polar glacier, the eyewitness of the climate change.**
  Author: D. Chinarro, J. L. Villarroel and J. A. Cuchi, C. Dominguez and A. Eraso (2012).
  Congress: International Geography Congress, IGC COLOGNE 2012, Germany, August 2012.
- **Climate change scenarios for temperature and precipitation in Aragón (Spain) (2013).**
  Authors: J. Ribalaygua and M. R. Pino and J. Pórtoles and E. Roldán and E. Gaitán and D. Chinarro and L. Torres
  Journal: Science of Total Environment.
- **A Study of the Relationship between Weather Variables and Electric Power Demand inside a Smart Grid/Smart World Framework (2012).**
  Authors: Luis Hernández, Carlos Baladrón, Javier M. Aguiar, Lorena Calavia, Belén Carro, Antonio Sánchez-Esguevillas, Diane J. Cook, David Chinarro and Jorge Gómez.
  Journal: Sensors 2012, 12, 11571-11591; doi:10.3390/s120911571. ISSN 1424-8220.

- **A Multi-Agent-System Architecture for Smart Grid Management and Forecasting of Energy Demand in Virtual Power Plants (2013).**
  Authors: Luis Hernández, Carlos Baladrón, Javier M. Aguiar, Belén Carro, Antonio Sánchez-Esguevillas, Jaime Lloret, David Chinarro, Jorge J. Gómez, Diane Cook.
  Journal = IEEE Communications, volume 51, No. 1, pages 106–114.
- **Modelos matemáticos aplicados al Acuífero de Fuenmayor (Sierra de Guara) (2010).**
  Authors: David Chinarro. Proceeding semminar: Aguas Subterráneas. IV Jornada. Ciclo sobre residuos. Published by Fundación Genes y Gentes (2010), page: 57–66.
- **Avances en localización, comunicaciones TTE y geofísica aplicables a cavidades de interésturístico (2010).**
  Authors: Ayuso, N; Bataller, V; Aisa, J; Cabrero, S; Chinarro, D; Cuchi, JA; Larma, R; Lera, F; Mediano, A; Molina, P; Muñoz, A; Tardioli, D; Sicignano, D; Viñals, V; Villarroel.
  Congress: III Congreso Español de Cuevas Turísticas (CUEVATUR 2010). Aracena (Huelva), 4–6 Noviembre, 2010. Published by J.J. Durán y F. Carrasco (Eds.), Cuevas: Patrimonio, Naturaleza, Cultura y Turismo, pp. 363–370, Madrid, Asociación de Cuevas Turísticas Españolas. ISBN: 978-84-614-4630-8. EAN: 9788461446308.
- **Sistemic Analysis of Karst Springs. Application to Fuenmayor spring (San Julián de Banzo, Huesca)**
  Authors: José Luis Villarroel and David Chinarro.
  Institution: Estudios Altoaragoneses, Diputación de Huesca. Report (2009–2010).

I lovingly dedicate this thesis to my wife, who always stood by me in all my difficult duties; to my children, to encourage them in the adventure of acquiring and sharing new skills; to my parents, who instilled me the endeavor in the study from an early age

Supervisors' Foreword

System engineering is typically defined as an art and science to create principles, identify methodologies, and establish models, with the purpose of deriving knowledge on controlling and managing the operation of complex-engineered systems. Similarly, a natural system can be categorized as complex, because of the difficulty in identifying and quantifying all information involved in its behavior, in terms of establishing all constraints and variables involved in the system, and accurately quantifying the temporal and spatial parameters. David Chinarro's thesis tackles fundamental questions posed in two natural systems, viz, a karst aquifer system and a glacier system, on the basis of transferability of system engineering methodology and from an exclusively hydrological outlook. This work has been developed in a high-tech research group, the I3A institute at the University of Zaragoza, where we carry out a set of projects by applying computing and communication technologies to groundwater, nivology, and geology. The guidance of this thesis by two mentors has allowed for it to be developed in a multidisciplinary way, due to complementary profiles from an engineer and a hydrogeologist.

Identification techniques, developed to represent typical engineering artificial systems through linear and nonlinear models, can be applied in the study of natural systems, where coupling effects between climate and hydrosphere occur. This thesis proposes enhanced methods to meet the new requirements in identification fields, innovating analysis and identification strategies to determine models that better represent the peculiarities of the two particular systems. One study specifically aims to estimate the mutual influence between precipitation and discharge of a karst aquifer, which is situated near Spanish Pyrenees mountains. Another, using the same data-driven models, deals with the implications of air temperature in a glacier melting, specifically, as it manifests itself in the stream drainage of Collins glacier, King George island, Antarctica. In this regard, special tools, such as those based on wavelet transform, have been considered in the preparing and analysis of time series, such as smoothing of signals, sampling frequency, coherence levels, and data abnormality detection. Through parametric and nonparametric identification processes, the author tries to identify models that best represent the internal dynamics of the system by iterative testings, where models are systematically checked against monitorized data, with an efficiency criterion given. The best obtained solution, in the assessment of results drawn from dealt cases, has been found among the model structures in blocks.

This thesis is meant to be a formal statement of engineering system identification methodology, mainly through nonlinear approaches, in the context of aquifer systems, which improves in many cases the results of karst hydrology. Remarkable results are derived from the characterization of Fuenmayor spring response and its correlation with precipitation, under the assumption of a linear system to be complemented with identification methods based on nonlinear techniques. Important findings such as a transfer function reduced to five parameters may describe properly the hydrologic behavior of this karst system under linear assumption. A Hammerstein–Wiener model approach presents a high efficiency value that exceeds the results obtained by the linear models using the same efficiency criterion. Likewise, approaches proposed for Antarctica's glaciology, through wavelet analysis and runoff models, scarcely appear in the literature, and can reveal essential information when it is not possible to clarify the all physics governing the system. The wavelet coherence density is used to estimate the boundaries of the Seasonal Effective Core (SEC), the period when the glacier discharge responds in a coherent way to temperature. For each year, an acceptably coherent sampling period is used as refined data for the identification process of the glacial system. Linear parametric identification was applied to each SEC. Nevertheless, it is demonstrated that nonlinearities present in glacier behavior are due to the phase change of water, a fact revealed by Hammerstein–Wiener structures used to define models with greater efficiency. Moreover, two different types of glacier dynamics have been discovered depending on the annual cycle and the SEC average temperature.

Therefore, system engineering techniques based on black-box identification, as a well-developed scientific procedure, are very useful tools to characterize the hydrological response of a karstic system discharge and to model the response of glacier stream to melting processes or climate variations.

Zaragoza, February 2014

José L. Villarroel
José A. Cuchí

Acknowledgments

It is very difficult to express in words the great gratitude I feel for who helped me to bring to fruition this work.

First, I would like to thank my two supervisors, Dr. Villarroel and Dr. Cuchí, because of their mentoring work, rich technical discussions and excellent guidance during the thesis development. Their wise advices and long experience have really contributed to the best findings of this work. Also, I would like to recognize the learnt lessons from lecturers of postgraduate and doctoral courses of University of Zaragoza, Group of Technologies in Hostile Environments (GTE) and Engineering Research Institute of Aragon (I3A).

Some individuals and entities have selflessly supported me and provided the necessary data to perform the exposed analysis. Dr. Adolfo Eraso and Dr. Carmen Dominguez, from GLACKMA foundation (Glaciers, CrioKarst and Environment), have supplied long-time series from glacier sensoring obtained under harsh conditions. Dr. Bartolome Andreo has outstandingly assessed and recommended the findings of this thesis, regarding it deserves an international dissemination. I must recognize the contribution of several Antartic bases: the Bellingshausen Russian base, former Jubany Station-Dallmann laboratory, Uruguayan Artigas base and Chilean Capitania del Puerto Bahía Fildes, for the meteorology data supply and logistical support. Also, thanks to authorities of *Natural Park of Sierra and Canyons of Guara* for granting access permissions to the Fuenmayor aquifer enclosure and its surrounding.

Gratefulness to managers and officials of San Jorge University, who have allowed me to participate and lead international research projects with a notorious impact on certain issues of the current thesis. Likewise, to some of my colleagues for their help and encouragement.

Acknowledgments

Contents

Symbols

Lists of symbols used in this paper with a brief description.

Symbol	Description
$\mu S/cm\, 25\,°\mathrm{C}$	Micro-Siemens per centimetre. Conductivity unit
θ	Parameters vector
$u_n(t)$	Discrete time series for the input of system
$y_n(t)$	Discrete time series for the output of system
$\hat{y_n}(t)$	Estimate value of $y_n(t)$
f*(x)	Complex conjugate of function f(x)
$\mathscr{F}\{u(t)\}$	Fourier transform of u(t)
$\widetilde{\mathscr{F}}\{u(t)\}$	Discrete Time Fourier Transform of u(t)
$\hat{f}(\omega)$	Fourier transform of the function f in the frequency domain
$\widetilde{f}(\tau,\omega)$	Wavelet transform of the function f in the time-frequency space
$\mathscr{G}_{\psi,s,\tau}\{u(t)\}$	Wavelet power spectrum of u(t) with wavelet family ψ
$\mathscr{H}\mathscr{W}_{f_{1,[nl]},f_{2,[nl]},f_{3,[nl]}}$	Hammerstein–Wiener model of u(t) and y(t)
$L^2(\mathbb{R})$	Set of all square integrable functions in the Hilbert space
$(f \otimes g)(t)$	Convolution of two function f(t) and g(t)
$(f * g)(t)$	Correlation of two function f(t) and g(t)
$\mathscr{L}\{y(t)\}$	Laplace transforms of y(t)
$\mathscr{W}_{\psi,s,\tau}\{u(t)\}$	Continuous Wavelet Transform of u(t) with wavelet family ψ
$\mathscr{P}_{\psi,s,\tau}\{u(t)\}$	Wavelet power spectrum of u(t) with wavelet family ψ
$\mathscr{C}_{\psi,s,\tau}\{u(t),y(t)\}$	Wavelet coherence spectrum
$\Gamma_{\psi}[f(t),g(t)]_{\tau}$	Coherence Average Function
σ_y^2	Variance of $y(n)$
C_y^k	Autocovariance of $y(n)$ with lag k
r_y	Autocorrelation of $y(n)$

Acronyms

Lists of abbreviations used in this chapter with a brief description.

AAO	Antarctic Oscillation
ACF	Autocorrelation function
ACVF	Autocovariance, or covariance of the signal
AIC	Akaike Information Criterion
ANN	Artificial neural networks
AR	Auto Regressive
ARIMA	Autoregressive integrated moving average
ARMA	Autoregressive-moving-average
ARMAX	ARMA with exogenous inputs model
BJ	Box–Jenkins Model
CAF	Coherence average function
CCM	Collins Coalescent Model
CWT	Continuous Wavelet Transform
CWS	Wavelet coherence spectrum
DDF	Degree-day factor
DDM	Data-driven model
DOF	Degree of Freedom
DSP	Digital Signal Processing
DTFT	Discrete Time Fourier Transform
DWT	Discrete Wavelet Transformation
EBM	Energy balance model
ECV	Essential Climate Variables
EOF	Empirical orthogonal functions
FIR	Finite Impulse Response
FFT	Fast Fourier Transform
FKS	Fuenmayor Karst System
GCM	Global circulation model
GLACKMA	Fundation for the study of glaciers, cryokarst and environment

GHG	Greenhouse gases
TIM	Temperature-index model
IIR	Infinite Impulse Response
ICSU	International Council for Science
IPCC	Intergovernmental Panel on Climate Change
KGI	King George Island
K–L	Kullback and Leibler information
LTI	Linear and Time Invariant
MIMO	Multiple Input–Multiple Output
MRA	Multiresolution analysis
MSLP	Mean Sea Level Pressure
NaN	Denomination for missing, or non-finite value
NLS	Non Linear Squares
NHS	Natural Hydrological system
OE	Output Error Model
ODE	Ordinary differential equations
PDF	Probability Density Function
PDE	Partial differential equations
PSD	Power spectral Density
SAM	Southern Annular Mode
SCAF	Smoothed Coherence Average Function
SISO	Single input-single output
STFT	Short-time Fourier Transform
SNR	Signal to noise ratio
TPH	Transition Period Head
UNEP	United Nations Environment Programme
UNESCO	United Nations Educational, Scientific and Cultural Organization
USGS	US Geological Survey
WFT	Windowed Fourier Transformation
WGMS	World Glacier Monitoring Service
WMO	World Meteorological Organitation
WPS	Wavelet power spectrum
XWS	Cross wavelet spectrum

Chapter 1
Introduction

All things are from water and all things are resolved into water.
Thales of Miletus (fl. c. 624 BC–c. 546 BC)

1.1 Justification

Water is an omnipresent substance on the planet, essential for all forms of life and present in diverse natural processes. The ubiquity, the mobility and the different states of water, make the study of hydrology complex. Concerning water, there are a huge variety of approaches, from scientific and professional perspectives. The boundaries between them can sometimes seem blurred, with large overlaps in the knowledge. Therefore, this introduction elucidates the core concepts that are the theoretical support of the tools applied in this dissertation. Also, it describes the motivation and establishes the objectives.

One of the most important phenomena in the Earth is the **hydrological cycle**. Water evaporates from seas, lakes, rivers, soils, ice and plants, and moves through the atmosphere to precipitate out as rain or snow. It forms, in diverse ways, streams, rivers, lakes, permeates soils, and contributes to the groundwater of aquifers. Examples of water necessity in biological processes are the plants that stretch their roots to draw on the soil moisture, which, along with minerals and the Sun's energy, produces the photosynthesis effect and the flora growth. The snow solidifies turning into ice sheets and glaciers-forms, which, depending on meteorological conditions, are melt to feed the streams and rivers, to finally deliver the water into oceans, waiting for the start of a new cycle.

In the hydrological cycle, the water continuous process, circulating throughout the Earth and its atmosphere, undergoes some incursions and vicissitudes as a result of anthropogenic or non-anthropogenic causes, all of which constitute so-called "climate change", manifested in different scenarios [1]. If all land ice melted away,

D. Chinarro, *System Engineering Applied to Fuenmayor Karst Aquifer (San Julián de Banzo, Huesca) and Collins Glacier (King George Island, Antarctica)*,
Springer Theses, DOI 10.1007/978-3-319-08858-7_1

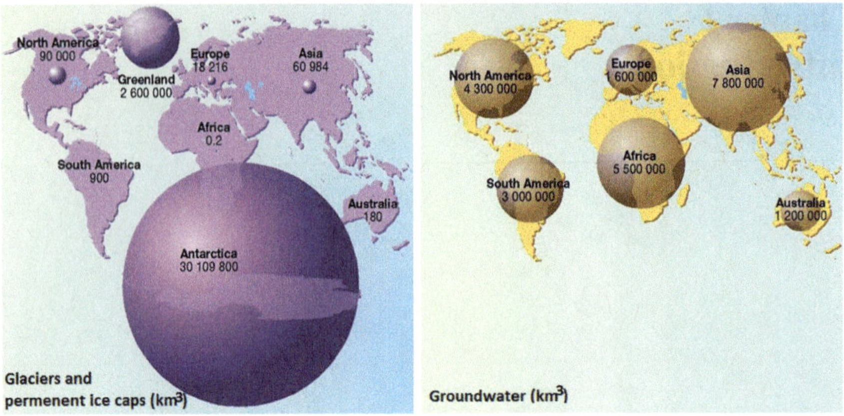

Fig. 1.1 Estimates referred to the volume of glaciers and ice caps (*left*), and groundwater (*right*). Both are the major topics to target mathematical models of this thesis. *Source* Rekacewicz [5]

the sea level would rise by almost 65 m. The contribution from ice sheets of Antarctica would be about 57 m and from Greenland 7 m [2].

The global water cycle has five main reservoirs that include the oceans (97.4 %), the atmosphere (0.0011 %), surface water (0.02 %), groundwater (0.59 %) and ice cap with glaciers (1.98 %) [3]. According to this report, surface water occupies a volume 100 times less than frozen water mass. About three quarters of total freshwater resources on Earth are stored as ice and mostly found in Antarctica (Fig. 1.1 Left). The groundwater reservoir, with a volume 30 times greater than surface water, has a special relevance in all continents for fresh water supplying (Fig. 1.1 Right).

Renewable and useable water resources in the world (renewable superficial runoff and aquifer recharge) are an estimated 42,785 km^3/year [3]. As a numerical example, in 1995, 3,800 km^3 were extracted for human use, 2,000 km^3 were consumed, and the rest was returned in much poorer quality. These fresh resources are renewable thanks to a natural cycle. If the current tendencies persist, world water withdrawals will likely reach 4,300 km^3 or even 5,200 km^3 in the year 2025. People around the world might soon be using between 61 and 74 % of all the available water resources [4]. Therefore, water is, and will be even more in the near future, a key issue for the economy everywhere. Water resources are scarce in most parts of the world. Awareness about the importance of sustainable management of water is growing throughout international organizations, committees to develop knowledge, and the current management of the waters in the world.

Karst system is a set of homogeneous or heterogeneous rocks, or geological formations with voids, pores, cracks, or fractures filled with water. Water flows from within the karst aquifer and emerges through springs. Karst landscapes are present in large parts of the world,—with approximately 30 % of the Earth's surface having a karst topography [6]. Most important karst areas are also often located in densely populated regions, and about 25 % of the world's total population are living off of karst aquifers [5]. Because karst areas constitute a substantial part of the world's lands, and are an essential resource of the freshwater for a great part of the world's population, there is a need to better understand the karst and improve the management of the water supply.

Glaciers and icecaps cover about 10 % of the world's landmass. According to the United States Geological Survey (USGS), 96 % of the world's frozen freshwater is at the South and North Poles (Fig. 1.1). These freshwater resources are not easily exploitable because most of them are far from human settlements, and the accessibility is not straightforward [3]. Nevertheless, subpolar glacier network can play the role of natural sensors of global warming evolution [7].

Hydrological models can be physical, conceptual (including lumped and distributed) and empirical, according to general terminology used in the hydrology literature. The empirical model applies mathematical equations that are not derived from physical processes in the catchment area, but from the analysis of time series data and statistical inferences. The technique also called data-driven modeling (DDM) is based on analysis of data from one system, seeking in particular connections between the system state variables (input and output variables) without explicit knowledge of physical behavior. The latest advances in hydroinformatics, an emerging field in the 21st century especially in data mining and computational intelligence running over a high performance architecture, have significantly contributed to progress on the empirical model.

This document is mainly focused on specific hydrological aspects, where factors involved in the water flow can determine the dynamics of the karst and glacier systems. It seems clear the necessity of applying proper and precise management models, based on advanced techniques in system identification, to natural systems, especially in systems with nonlinear features and extreme events, e.g., long droughts and heavy downpour in a spring area, or outbursts in the glacier discharge during transitional seasons.

1.2 Motivation

International experts revealed in April 2010, at the V International Symposium on Karst, held in Malaga, the importance of research on and knowledge of karst aquifers. They agree on the need to plan actions for enhancing the knowledge and necessary control over the use and management of groundwater reserves and sustainable development. Among others, they highlighted this idea: *groundwater resent much less than the surface waters when a prolonged drought due to the large storage capacity they have* [8]. Starting from linear models already validated, the nonlinear system hypothesis to identify the karst system has been set as an aim to improve the knowledge of the system.

Water in solid form, such as snow or ice, is a major field on hydrological sciences. On an empirical foundation developed over decades, the GLACKMA project [9] means an exhaustive contribution to the polar glaciology. Domínguez et al. [10] show some correlations between glacier discharge and climate variations. For this purpose, since 2001 eight stations have been installed as Pilot Experimental Watersheds at different latitudes in the glaciers of both polar circles, with continuous recording of glacier discharge at hourly intervals, and they offer the possibility to be treated by other approaches.

The premises above, along with the valuable studies of predecessors—those who have tackled research on the karst spring in San Julian Fuenmayor Banzo (Huesca) [11] as well as those who have monitorized and analyzed the Collins glacier in Antarctida [12, 13]—and together with my personal involvement and awareness of the environment, are the principles and motivation to develop a treatise along in terms described below.

The motivation for developing this PhD thesis is boosted by

- My former and lasting fondness for system identification and mathematical modeling from many years ago, as evidenced my participation in a major research project on modeling and simulation [14].
- A challenge in demonstrating my capabilities developed over many year of research experience as a science computer lecturer, plus fundamentals acquired in the Computing Postgraduate and PhD Master of University of Zaragoza.
- The essential support and knowledge of experts and scholars who have encouraged me to tackle the posed problems in the karst spring of Guara Mountains and the glacier of Antarctica.
- My ability acquired by management and leadership of large international projects on computer science, software engineering, process automation and mathematical models, in the fields of industrial automation, sensoring, distributed energy microgrids, environment, and climate change assessment.
- An endeavor and personal involvement with sustainability in natural resource consumption and my sensitive awareness of the state of the environment which will be inherited by the next generations.

1.3 Basic Concepts

System is a collection of interrelated elements that form a whole and with general properties of the whole rather than of the individual elements [15]. The word "system" derives from the Greek "synhistanai" ($\sigma\,\nu\sigma\,\tau\,\eta\mu\alpha$) which means "to place together". In the ample sense, the term "system" may mean an engineered system, a natural system, a social system, or all three.

Systems science provides methods to address complex problems, which enable researchers to examine the dynamic interrelationships of variables at multiple levels of analysis, and study the impact on the behavior of the system as a whole over time. Moreover, simulation modeling can be used to generate forecasting, allowing decision makers to simulate the impact of alternative solutions before carrying them out [16].

Checkland classified four **types of systems** about the real world after considering Boulding's hierarchy and Jordan's taxonomy. The four systems include *natural system, engineered system, abstract system* and *human activity system* [17]. Hence, the concept of system also serves to identify those manifestations of natural phenomena and processes with complex relationships among them. Anything that can belong to the Earth domain is a natural system, component of a natural system, or an aggregate formed by natural systems. In searching for the appropriate method and methodology for specific situations, according to Zexian and Xuhui [17] methods concerned with different Checkland's classes can be interchangeable. Hence, techniques used in engineered system could be useful in natural systems.

Mathematical models are formal statements or equations to express the relationship between system inputs, outputs and operations, to accomplish hydrological simulations. Then, mathematical models, as reasonable representations of the system, provide solution approaches to a wide spectrum of cases: for example, in streamflow or spring forecasting, recovering values in missing data, quantifying the impact over land changes planning, designing and managing water resources, better understanding the hydrological processes, among others.

The term **Systems Engineering** is a generic term that describes the application of structured engineering methodologies to the design of complex systems which require high dynamic performance.

Methods of system engineering are interdisciplinary tools to emulate complex systems; therefore multidisciplinary applications are required. These methods involve elements of mathematics, physics and computing as well as techniques of analysis and control, to provide solutions in many different fields; e.g., electronics, industrial mechanics, distributed computing, power, communication networks, manufacturing, logistics, artificial vision, robotics, transportation, chemical processes, medical and biological systems, environmental systems, and bioprocesses.

Since **general theory of systems** attempts an integrative methodology for the treatment of scientific problems [18], natural systems can gain the transferability of models from different scientific continents.

The study of **natural systems** in this thesis leverages the systems engineering tools following the system identification techniques (Fig. 1.2). For that reason, karst and glacier systems, from a data-driven analysis, are going to be characterized and identified by engineering methods essentially from a non-linearity approach.

From the perspective of systems engineering, the karst or glacier system can be assumed as a *black box* device with behavior dictated by physical laws, which are not fully known, but observations of the inputs and outputs can be useful in formulating a specific idea of the system dynamics.

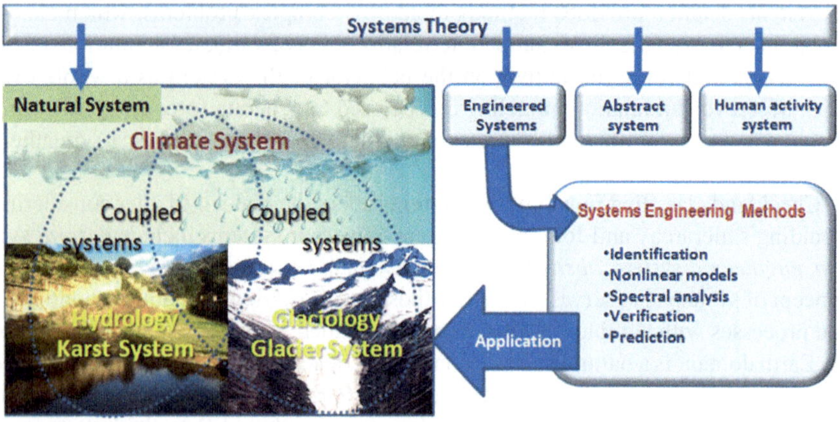

Fig. 1.2 Checkland's taxonomy [17]: natural system, designed physical system (engineered system), designed abstract system and human activity system. Climate and karst (or glacier) are coupled systems in the context of system theory. Natural systems can leverage the tools from systems engineering

1.4 Hypothesis

System identification researchers should make decisions based on the following perspectives to cope with characterizing or predicting the behavior of a system based on recorded data:

- How can they analyze the signals to obtain as much information as possible about the system?
- How can they best use the information in the observed data to calculate a model with the same properties and behavior as the real system?
- How can they know if the model is any good and how can they rely on it for simulation, design or prediction purposes?

In the case of karst systems and glacier systems, these three questions are kept open to compelling answers, though already there are many excellent approaches.

As a starting point of this thesis to formulate the targets, and carry out a formal study applying the scientific method at each stage, I had to ask myself some questions on observed facts to pose the problem properly, such as:

- What is the parametric interdependence between input and output time series?
- How can we unveil features in the signals of an aquifer that otherwise would remain hidden by methods heretofore known?
- Why, in the same recharge area, with the same precipitation, does an aquifer present a different flow regime than another?
- Can I predict the discharge effects next time it rains in the recharge area of Ciano Polje (San Julian de Banzo)?

- Does the complexity of nonlinear models in its theoretical structure and its computational implementation compensate improvements over linear models?
- How can the interaction between climate and hydrological cycle be quantified by correlation analysis?
- To what extent can a linear model emulate the dynamic behavior of a glacier?
- Can be calculated the bounds of the active period in a glacier?
- Which could be the suitable nonlinear structure for the annual cycle evolution of a glacier?
- How can the glacier dynamics be classified by the features of the glacier in the active periods?
- Would it be possible to achieve a generalization of the performance of a particular glacier in terms of a global model?
- To what degree can a glacier play the role of change climate sensor?

1.5 Aims of the Thesis

The primary objective of this thesis can be stated as
Applications of linear and nonlinear methods of system engineering to natural systems, namely the karst aquifer and the Antarctic glacier, in order to reveal the relationship between variables through spectral analysis, searching for the best model for simulation and prediction, and contributing with advanced solutions that help optimize the management of a karst aquifer and characterize the hydrological dynamics of a glacier.

The scope of this study has been the design and verification of models based on some linear techniques, with the support of spectral analysis based on wavelets techniques. Likewise, a set of selected nonlinear model structures has been used in the identification processes, focused on two natural scenarios previously chosen. One of them is the karst aquifer of Fuenmayor (San Julian de Banzo, Huesca, Spain), for the treatment of data collected from the years 2002 to 2005, and another is on Collins Glacier (King George Island, Antarctica) with data in the period 2001–2011.

1.6 Precedent Context

The Group of Technologies in Hostile Environments (GTE) from the University of Zaragoza (Spain) was founded in 1997, and its work is mainly oriented towards natural and/or hostile environments, such as high mountains, polar zones, canyons and underground environments (caves, mines, tunnels). The GTE, a multidisciplinary group integrated by researchers and lecturers, provides the proper background to develop a thesis like this. One of its research lines deals with the study of karst springs, mainly through a new system identification techniques, with results gathered in papers that come from over 10 years ago.

Glackma Foundation promotes scientific research in the polar regions. Its researchers have been visiting both poles since 1985, almost each year. They have registered long time series related to glaciers discharge in Arctic and Antarctic glaciers, to study the evolution of global warming, using glaciers as natural sensors. Glackma Foundation has provided the data and the fundamentals to carry out the part of this work about glaciers.

1.7 Outline of the Thesis

After outlining the objectives, justification and fundamental assumptions of this work in the current chapter, this thesis is organized according to the following scheme:

Chapter 2 is a description of the main tools from system engineering that will be applied in the cases under study. It presents the theoretical basis to understand the fundamental of system identification techniques. It starts with classical analysis techniques and an overview of some existing methods for linear and nonlinear systems, with an introduction to an important block structured nonlinear system. The chapter also deals with particulary wavelets techniques, showing the powerful advantages of these mathematical tools that are going to be used extensively in the experimental cases considered. The last sections explain parametric and non parametric identification and finally the nonlinear identification is presented.

Chapter 3 is devoted to the presentation of two natural systems to understand details of the conceptual models, identify the characteristics of the environment that condition the model design, and determine the scope of the problem to be worked out. The first review is focused on the karst system scenario, with a description of the karst hydrology and the input/output variables of the problem. The second one describes the Antarctic environment, especially the melting processes in the glaciers. In order to understand weather and discharge correlation models, some fundamentals should be set. The chapter highlights the natural system features in both systems, and reviews the precedent models in the literature, in order to address the analysis and identification in the next chapters.

Chapter 4 begins with the description of the karst aquifer at the Fuenmayor site. It follows analysis and system identification results after applying Linear time-invariant system theory, then nonlinear methods based on wavelet transform are applied to rainfall and aquifer discharge, following some chosen techniques from Chap. 2 to obtain a system nonlinear model.

Chapter 5 is the study of Collins Glacier in Antarctica, drawing on fundamental techniques that have already been depicted in Chap. 2. It details some levels of coherence to work out the suitable seasonal events for several years. Procedures based on linear methods have been tested, along with nonlinear tools, especially block models and analysis based on wavelet transform.

Chapter 6 summarizes the most important conclusions of the results that have been reached throughout this thesis. It highlights most significant research contributions and possible extensions of the findings. Also, it discusses suggestions for further developments and improvements.

Finally, complementary information are contained in the appendices. Appendix A contains examples of analysis results in the form of illustrations, which because of their complementary character, or over sizing, are not included in the text of the corresponding chapters. Appendix B is a glossary of the main terms defined according to the meaning given in the thesis context.

References

1. Ribalaygua J, Pino M, Pórtoles, J, Roldán E, Gaitán E, Chinarro D, Torres, L (2013) Climate change scenarios for temperature and precipitation in Aragón (Spain). Sci Total Environ 463–464:1015–1030
2. Reinwarth O, Stäblein G (1972) Die kryosphäre—das eis der erdeund seine untersuchung. Würzburger Geographische Arbeiten 36:71
3. Shiklomanov I (1998) World water resources. A new appraisal and assessment for the 21st century. Technical report, United Nations Educational, Scientific and Cultural Organisation (UNESCO), Paris
4. Valero A, Botero E, Serra L (2002) The world's renewable water resources and ice sheets. In: Proceedings of the conference on sustainable development of energy, water and environmental systems. Dubrovnik, UNESCO, Croatia, 2–7 June 2002
5. Rekacewicz P (2009) Freshwater resources: volume by continent. In: Vital water graphics 2. GRID-Arendal, in collaboration of UNEP. Accessed 27 Sep 2013
6. Milanovic PT (2009) Water resources engineering in KARST. Publisher
7. Eraso A, Domínguez M (2006) Subpolar glaciers network as natural sensors of global warming evolution. Ukranian Antarct J 4–5:272–277
8. Andreo B, Carrasco F, Durán J, LaMoreaux J (2010) Advances in research in karst media, vol 1. Springer, Heidelberg
9. Eraso A Domínguez C (2005) GLACKMA (Glaciers, Cryokarst and Evironment). http://www.glackma.es/. Accesed 19 July 2012
10. Domínguez M, Rodriguez-Puebla C, Encinas AH, Visus I, Eraso A (2010) Glacier discharge and climate variations. In: European geoscience union. General assembly 2010. Geophysical Research Abstracts, vol 12, pp 555–567
11. Cuchí J, Villarroel J, Carceller T (2006) Comportamiento del manantial de Fuenmayor, (San Julián de Banzo, Huesca) durante la sequía de 2005. In: Congreso Internacional sobre: El agua subterránea en los países mediterráneos, AQUAINMED06, pp 24–28
12. Domínguez M, Eraso A (2009) Substantial changes happened during the last years in the icecap of King George. Technical report, PANGAEA—Publishing network for geoscientific & environmental data, Insular Antarctica
13. Eraso A, Domínguez C (2010) Sampling data at CPE-KG-62°S. Collins glacier, GLACKMA
14. Romay I, Calvete H, Chinarro D, Lasala P, Moreno JM (1988) Modelos matemáticos de decisión y simulación aplicados a la táctica. Resumen. In: Fomento a la Investigación. Memoria 1988, vol 1, p 305. Diputación General de Aragón, Zaragoza
15. Bertalanffy LV (1968) General system theory; foundations, development, applications. George Braziller, New York

16. Sterman JD (1994) Learning in and about complex systems. Syst Dyn Rev 10(2–3):291–330
17. Zexian Y, Xuhui Y (2010) A revolution in the field of systems thinking– a review of checkland's system thinking. Syst Res Behav Sci 27(2):140–155
18. von Bertalanffy L (1950) An outline of general system theory. Br J Philos Sci 1(2):134–165

Chapter 2
System Identification Techniques

> *An intellect knowing at any given instant of time, all forces acting in nature, as well as the momentary positions of all things of which the universe consist, would be able to comprehend the motions of the largest bodies of the world and those of the smallest atoms in one single formula, provided it were sufficiently powerful to subject all the data to analysis.*
>
> Pierre Simon Laplace (1749–1827)

2.1 Introduction to System Identification

MODELING is the abstraction of a real process to characterize its behavior. Scientific modeling aims to enhance the investigation of phenomena in order to reveal and better understand cause-effect relationships [1]. The model definition given by Eykhoff [2] introduces the concept of "essential aspects": "... *[model] is a simplified representation of the essential aspects of an existing system (or a system to be constructed), which presents the knowledge of the system in a usable form*".

The set of processes in a system determines the behavior of the system. Every process is determined by its physical and chemical properties, which are not always easily known. A model tries to emulate the 'essential aspects' of the system behavior, simplified by choosing the most significant properties. So, **modeling techniques** can be classified as:

- *a priori* modeling, *white-box* or morphological modeling, by making simple experiments to inquire into the physical or chemical laws involved.
- *a posteriori* modeling or *black-box* modeling, by building a model based only on data (data-driven) without having previous knowledge of the system. The model describes how the outputs depend on the inputs, not how the system actually is, and characterizes the system dynamics (delays, speed, oscillations, and others), though the physical interpretation of the results is not straightforward.

D. Chinarro, *System Engineering Applied to Fuenmayor Karst Aquifer (San Julián de Banzo, Huesca) and Collins Glacier (King George Island, Antarctica)*, Springer Theses, DOI 10.1007/978-3-319-08858-7_2

- *grey box* modeling is an intermediate technique when peculiarities of internal laws are not entirely known, so it is based on both insight into the system and on experimental data analysis.

System identification tries to estimate a *black* or *grey* model of a dynamic system based on observing input-output from experimental data. Zadeh (1962) defined *system identification* as: "... *the determination on the basis of input and output, of a system (model) within a specified class systems (models), to which the system under test is equivalent (in terms of a criterion)*".

The availability and reliability of the design techniques of system identification did expand the application fields beyond the scope of industrial applications. As a result, system identification models have been applied in other diverse fields, for example, economy, environment, biology, psychology, biomedical research, hydrology, and glaciology. The identification problem requires a set of model structures, a validation criterion and an aim [3]. Criteria and models will be presented over the course of this chapter. Some examples of identification aims could be listed here:

- To design control strategies for a particular system (e.g., in optimizing an electrical microgrid operation).
- To analyze the properties of the system (e.g., quantity rates in a medication reaction).
- To forecast the evolution of the system (e.g., future climate prediction according a IPCC downscaling model)
- To identify hidden factors influencing a system (e.g., sun spots in the karst spring).
- To improve the internal knowledge of the system (e.g., the delay in the aquifer discharge with respect to precipitation events).
- To identify the interaction between coupled systems (e.g., climate and glaciers).

The objective of this chapter
This chapter surveys the main approaches to identification and analysis of systems and their theoretical basis, in order to set the methodological scene for the study of the experimental cases described in later chapters, especially highlighting the procedures that mostly have been implemented in natural systems.

Outline of this chapter
The first Sect. 2.2 of this chapter deals with problems in the acquisition of data from sensors including the sampling chosen (2.2.1) and treatment of outliers (2.2.2) to reach a sufficient time series quality for the subsequent treatment of the information.

Section 2.3 is a review of the classical time series as the first approach to understand the underlying mechanism in the system. Two scenarios are presented: time domain analysis (2.3.1) with autocorrelation structures, and frequency domain analysis (2.3.2) with an explanation of concepts like Fourier transform and frequency spectrum.

Although wavelet techniques are included in the frequency domain issues, a special section is dedicated to wavelet techniques (2.4), because of its prominent application in the practical cases. The section contains a description of mother wavelets, detailed clarification of wavelet transform (2.4.2), the process of estimation of the wavelet power spectrum (2.4.3), and some discussions on using these techniques (2.4.4).

A review of model structures (2.5) is necessary for the identification process to select suitable, identifiable model structures. Some linear time-invariant models are described in Sect. 2.5.1. Nonlinear models (2.5.3) go through Volterra series (2.5.3.1) and Hammerstein-Wiener models (2.5.3.2).

Section 2.6 deals with several identification techniques, starting with the elucidation of the posed problem (2.6.1) and a simple literature overview (2.6.2).

Parametric identification (2.8) relies on a model previously defined by a set of parameters that must be calculated to accomplish a given quality criteria. Linear techniques are given in (2.8.1) with some advice on selection and verification criteria of the models (2.8.2).

Nonparametric identification methods are described in 2.7, starting with methods in the frequency domain (2.7.2), including classical spectral analysis (2.7.2.1), going on to wavelet cross spectrum (2.7.2.2), and finishing with wavelet coherence fundamentals (2.7.2.3). In the time domain Sect. (2.7.1) there are methods for system identification, including cross correlation (2.7.1.1) and impulse response (2.7.1.2).

A final section provides an overview of nonlinear identification techniques (2.9) which serves as a tour of nonlinear parametric identification (2.9.1) remarking on the main issues of Volterra identification (2.9.1.1) and Hammerstein-Wiener identification (2.9.1.2). Additionally, methods pertaining to nonlinear nonparametric identification are mentioned (2.9).

Finally, in the conclusion Sect. 2.10, important assertions related to this chapter are offered.

2.2 Time Series

To capture critical information about the processes to be investigated, field data are achieved through the sensor network. Process variables should be sampled for a duration and sampling frequency enough to obtain those quality time series that the analysis requires. The sequence of observations on one variable y(t), t ∈ T (T is the discrete times domain), is called *time series*. The observation are usually equally spaced and indexed by integers (t = 1,…,n) where n indicates the number of observations. The main objective of time series analysis is to get mathematical inferences from the sample data obtained from field sensors.

A lot of problems can be solved by time series as: prediction (e.g., future weather precipitation), identification or abnormal peaks (e.g., outbursts in a glacier discharge), trends (e.g., sea global warming in the last century), etc.

Before performing the analysis of the time series, some issues should be addressed on the form of achieving raw data, check the integrity and reliability of preparing procedures; i.e., the sampling method, the outliers detection, and the lost data recovery.

2.2.1 Sampling Period

In many monitoring applications, there are some problem related to sensor devices to sample environment variables, mainly concerning with computing data as the memory size, processing capability and power supply. Sampling is the process by which continuous time signals—such as air temperature or water levels—are turned into discrete time signals. About batteries, if the sampling frequency is set too high, then the energy consumption would be so high that the sensor battery would be depleted too soon. To avoid this problem, the sampling frequency can be reduced, but this is not always possible. The Nyquist-Shannon sampling theorem states that if a function y(t) contains no frequencies higher than ω_N, it is completely determinable by a sampling process of frequency 2ω. So, the sampling frequency ω_S should be bounded according the Eq. 2.1:

$$2\omega = \omega_N < \omega_S < \omega_C \tag{2.1}$$

where ω_N is the Nyquist frequency and ω_C is the critical frequency for the sensor battery duration. On this matter, Alippi et al. [4] presented an adaptive sampling algorithm for effective energy management in wireless sensor networks. Some of these sensors require computing capability to integrate a distributed artificial intelligence which offers a wide range of possibilities for the operation, automation and control of different systems [5].

The Nyquist-Shannon criterion provides some clearly stated bounds, but it does not assure that a system can operate right at the Nyquist rate. So, some practical caveats should be taken in mind in designing a sampling process. The Nyquist criterion assumes: a sampling is regular (low noise); the value obtained for each sample should have an infinite precision (issue never happens in sampling natural system); there can be no component in the original signal higher than those correspond to half the sampling frequency; and the criterion does not take in account if (so prevalent in system identification) the signal is going to be processed (modified) after sampling [6].

2.2.2 *Outliers*

Before performing the analysis of the time series, a preparation phase is carried out to check the integrity and reliability of measurements. **Outliers** are the discordant and unexpected values that can appear in a time series, which Hawkins [7] defined as: "*the observation that deviates so much from other observations as to arouse suspicion that it was generated by a different mechanism*". Barnett and Lewis [8] named outliers as the data deviated markedly from other members of the sample in which it occurs.

Then, the goal of the outliers identification is the location of suspicious values to removed or recovered by a more suitable value, in order to avoid contamination and distortion of the underlying probability distribution. Really, it is hard to tell whether a value is an outlier or true data. So, outliers not always must be considered an erroneous data to be rejected. Outliers can be values produced by natural dynamics of the processes to study, but there are outliers that can be due to external factors of the system, as sensor device failure, mistake in entering data, malfunctioning of the datalogger, break in the communication, etc. [9].

Statistics can calculate if the probability of the candidate observation is small or is enough far from the rest, to be considered as an outlier. Among the distinct methods to detect and correct outliers, the Rosner criterion is well known [10]. A generalized extreme studentized deviate (ESD) is the fundamentals in Rosner criterion to detect from 1 to k outliers in a data set that follows an approximately normal distribution. This procedure handle the error both under the hypothesis of no outliers and under the alternative hypotheses of 1, 2,..., k−1 outliers. Rosner's method detects and removes multiple outliers in a single step.

Under the hypothesis that an excessive deviation of the expected spectrum of a signal is interpreted as the possible presence of outliers, a new test of outliers has been proposed and applied by Chinarro et al. [11] to raw data of Fuenmayor spring. It is the Wavelet-Rosner test, as an extension of Rosner test in the frequency domain (Sect. 2.4.2.4).

2.3 Classical Time Series Analysis

Time series analysis definition was given by Tukey [12] : "*Time series analysis consists of all the techniques that, when applied to time series data, yield, at least sometimes, either insight or knowledge, and everything that helps us choose or understand these procedures*". Thus, a time series can unveil some concealed information about the system, as periodicity, outliers and trends, using typical statistics estimators.

Univariate analysis methods characterize the structure of an individual time series—simple analysis—while *bivariate methods* study the relationship between two different time series—cross analysis.

Under the concept "classical time series analysis" two approaches are studied. One is the analysis in the *time domain* based on concepts like autocorrelation and autoregression. Another is the analysis developed in the *frequency domain* that deals with spectral analysis.

2.3.1 Time Domain Analysis

The time series y(t) is a discrete signal expressed as $\{y(kT)\}$ and $\{x(kT)\}$ that is a sequence obtained from the output (analogously x(t) for the input) of a system at sampling instants ($t_k = kT; k = 0, 1, 2, \ldots, N$), where T is sampling interval and N the number of samples.

Autocorrelation refers to the correlation of a time series with its own past and future values. In order to determine the expression for autocorrelation, the following functions over a discrete time series $y(kT) = \{y_1, y_2, \ldots, y_N\}$ are defined below.

(a) Variance:

$$\sigma_y^2 = \frac{1}{N} \sum_{k=0}^{N} (y(kT) - \mu)^2 \tag{2.2}$$

where μ is the mean of N elements in the time series $y(kT)$.

(b) Autocovariance with lag = τ:

$$C_y^\tau = \frac{1}{N} \sum_{k=0}^{N-\tau} [(y(kT) - \mu)(y(k+\tau)T - \mu)] \tag{2.3}$$

(c) Autocorrelation as the rate between the variance and the autocovariance:

$$R_y^N = \frac{C_y^\tau}{\sigma_y^2}; \quad \text{for any } \tau : R_y^N(\tau) = \frac{C_y(\tau)}{\sigma_y^2} \tag{2.4}$$

Coefficient R_y^N at lag τ given in Eq. 2.4 defines the autocovariance (Eq. 2.3) normalized by the variance (Eq. 2.2) [13]. Positive autocorrelation might be considered a specific form of 'persistence' of events, and time series is better predictable because future values depend on current and past values.

2.3.2 Frequency Domain Analysis

A useful tool to characterize signals is the power density spectrum, due to spectral analysis is concerned with estimating the relative importance of different frequency bands in a signal. The spectrum from a finite-length sequence of samples $y_N(kT)$, with (K = 0,1,...,N), is given by Eq. 2.5:

$$\phi_y^N(n) = \frac{1}{N}|Y_N(n)|^2 \tag{2.5}$$

where $Y_N(k)$ is the Discrete Fourier Transform (DFT) of the time series $y_N(kt)$ and calculated by:

$$\mathscr{F}\{y_N(kT)\} = Y_N(n) = \sum_{k=0}^{N-1} y_N(kT)\, e^{-(i2\pi nk/N)}$$

As $Y_N(n)$ is a discrete time signal, the frequency variable is also discrete $\omega = n\frac{2\pi}{N}$. Each $Y_N(n)$ is a complex number that encodes both amplitude and phase of a sinusoidal component of function $y_N(kT)$. The sinusoid frequency is k/N cycles per sample. $\phi_y^N(n)$ is called the *periodogram* of $y_N(kT)$. Periodogram analysis was proposed by Stokes [14] and applied by Schuster [15] to analyze sunspot data.

However, this estimate is very fluctuating and only gives a rough picture of the power density spectrum. See Wellstead [16] or Ljung and Glad [3] for an explanation of this problem. A first solution is to average a number of periodograms calculated over different segments of the full signal sequence. It is called the Welch's method [17]. The power density spectrum is the Fourier transform of the autocorrelation, as expressed in Eq. 2.6. See, for example, Ljung and Glover [18] or Ljung and Glad [3] for the basis of this estimate.

$$\phi_y^N(n) = \sum_{k=-N+1}^{N-1} R_y^N(k) e^{-(i2\pi nk/N)} \tag{2.6}$$

The most extended method to power density spectrum is the Blackman-Tukey procedure [19]. The idea is to smooth the periodogram by averaging over a number of neighboring frequencies using a windowed technique. For this last technique, another estimate of the power density spectrum is normally used:

$$\phi_y^N(n, W) = \sum_{k=-M+1}^{M-1} W(k) R_y^N(k) e^{-(i2\pi nk/N)} \tag{2.7}$$

where $W(k)$ is called the lag window and M is the window size that should be small compared with N. There are several window functions, but the most common used in spectral analysis is the Hamming window [3]. In hydrology, it is frequently necessary to numerically evaluate the real part of the *one-sided* Fourier transform of of autocorrelation. The expression of the spectral density function of a signal is given by:

$$S_f = 2\left[1 + 2\sum_{k=1}^{N} W(k)R_y^N(k)\cos 2\pi f k\right] \tag{2.8}$$

where W(k) is the lag window, commonly a Hamming window in spectral analysis.

2.4 Wavelet Transform Techniques

Wavelet techniques supply solutions to the time-scale analysis by decomposing a signal into a superposition of scaled and shifted versions of an original wavelet (also called mother) with specific properties, as a fast-decaying oscillating function. During the last 15 years, great strides in the development of the theory of wavelets have made. The search continues in new application areas and theoretical approaches. The literature devoted to wavelets is very voluminous, so a strict selection of fundamentals is treated in the next sections.

Wavelet theory emerges in the mid-1980s by Grossman and Morlet [20] who apply Gabor wavelets to model echo signals for underground oil prospecting. Jean Morlet was prospecting oil for the Elf-Aquitaine company, sending pulses to underground and analyzing their echoes by Fourier transforms; giving that, the high frequencies of the echoes correspond to thin layers and the low frequencies to the thicker ones. Nevertheless, he found that a lot of reflected signals corresponding to the different layers, were interfering with each other, and he can not separate the required ones. If he selects an extremely small windows (Windowed Fourier Transformation (WFT) [21]) to analyze only signals in high frequencies, all information about low frequencies would be lost. Morlet had to manage the problem in a different way. Instead of fixing the size of the window, he kept constant number of oscillations in the window and varied the width of the window, by stretching or compressing. This made possible to decompose signals simultaneously by time and frequency, giving useful information: *what frequency and when it is produced.* The wavelet engineering was born [22].

In a **historical overview** (Table 2.1), the wavelet theory starts from methods that have been essential in the development of engineering during almost two centuries. Haar [23] introduced the first compactly supported family of functions. After a long time elapse without contributions, Gabor [21] introduces a family of non-orthogonal wavelets with two components, a complex sinusoidal carrier and a Gaussian envelop. A wavelet filter banks for decomposition and reconstruction of a signal were introduced by Esteban and Galand [24], although Crochiere et al. [25] roughly introduce the same idea in speech acoustics, called sub-band coding.

The literature rapidly spread out and wavelet analysis is now used extensively in physics, geophysics [26], economy [27], epidemiology, neuroscience, signal processing Ricker [28], hydroclimatology [29], oceanography [30], hydrogeology [31], hydrology [32], electricity demand [11, 33], remote sensing data [34], computing complex problem as Maxwell's curl equations [35], and other fields.

Table 2.1 History of wavelets

Year	Authors	Facts
1807	J.B. Fourier	Any periodic function can be expressed as an infinite sum of sine and cosine waves of different frequencies. His ideas faced much criticism from Lagrange, Legendre and Laplace for lack of mathematical rigor and generality, and his papers were denied to be published, until 15 years later
1909	A. Haar	He discovers a "base" of functions that are now recognized as the first wavelets. They consist of a short positive pulse followed by a short negative pulse
1930	J. Littlewood and R. Paley	Local information on a wave, as the duration of a pulse of energy can be recovered by grouping the terms of its Fourier series in "octaves"
1946	D. Gabor	He adapted the Fourier Transform to analyze only a small section of the signal at a time. Gabor's adaptation, called Short-Time Fourier Transform (STFT), decomposes a signal into a two-dimensional function of time and frequency. The result is a "packets time-frequency" or "frequency Gabor"
1960	A. Calderon	He provides a mathematical formula that allows mathematicians subsequently recover a signal from its wavelet expansion
1976	D. Esteban and C. Galand	They recognize the subband coding, a way of encoding digital transmissions for telephone
1981	J. Morlet	He discovers a way to decompose the seismic signals where called *wavelets of constant shape*. Ask for help to Alex Grossmann, quantum physicist, to show that the method works
1982	E. Adelson and P. Burt	They developed the "pyramidal algorithm" for image compression
1984	Morlet and Grossmann	They first introduced the term "wavelet" in mathematical language
1985	Y. Meyer	Before 1985, a lot of researchers thought that there was no orthogonal wavelet except Haar wavelet. Yves Meyer constructed the second orthogonal wavelet called Meyer wavelet (soft orthogonal wavelets)
1986	S. Mallat	Shows that the Haar basis, the Littlewood-Paley octaves, frequencies and filters Gabor, and subband from Galand and Esteban, are all related to algorithms based on wavelets
1987	I. Daubechies	She constructs the first smooth orthogonal wavelets with a solid foundation and systematical method, to be used as a practical tool that any scientist can easily program and operate
1988	S. Mallat and Y. Meyer	Mallat and Meyer proposed the concept of multiresolution
1990	D. Donoho and I. Johnstone	They used wavelets to "remove noise" of images, making them even sharper than the originals
1992	T. Hopper, J. Bradley and C. Brislawn	They developed a method based on wavelet to compress its huge database of fingerprints that were applied in Criminal Information Services of FBI

(continued)

Table 2.1 continued

Year	Authors	Facts
1995	Pixar Studios	Presentation of film Toy Story 2, where some forms are provided by subdivision surfaces, a technique mathematically related to wavelets
1996	W. Sweldens	He introduced a new technique, the so-called lifting scheme, which became the basic tool of second generation wavelets
1999	The International Standards Organization	Published a new digital image compression called JPEG-2000. The new standard uses wavelet to compress image files in a ratio of 1:200, without appreciable loss in image quality

Source Daubechies et al. [36] and information gathered from wavelet literature

2.4.1 Wavelet Function

The wavelet transformation needs a basis function—like Fourier transform requires sinusoid basis function, though in wavelets, there are a wide range of basis function families. Principles and conditions of this basic functions are related below.

$L^2(\mathbb{R}^d)$ is a Hilbert space and denotes the set of square integrable functions, i.e., the set of functions defined on the real line such that $\int_{-\infty}^{\infty} |x(t)|^2 dt < \infty$. Since this integral is usually referred to as the energy of the function x, this space is also known as the *space of functions with finite energy*. Therefore, $L^2(\mathbb{R}^d)$ has an inner product $\langle x, y\rangle = \int_{-\infty}^{\infty} x^*(t) \cdot y(t)dt$, and an associated norm $\|x\| = \langle x, y\rangle^{1/2}$.

The term wavelets refers to a set of small waves formed by dilations and translation of a single function $\psi(t)$ which should be square integrable over the range of real time of space $L^2(\mathbb{R})$. Table 2.2 summarizes the requirements to be met by a wavelet [37].

The function $\psi(t)$ is called "mother wavelet" or "basic wavelet" while the dilated and translated functions derived from the "mother wavelet" are called "daughter wavelets" or simply "wavelets" (Fig. 2.1). These daughter wavelets have the same shape as their mother wavelet. Their amplitude should rapidly decay away from the center of the wave in both time and frequent domains. The functional relationship between daughter $\psi_{s,\tau}(t)$ and mother $\psi(t)$, in the scale s and displacement τ, is

Table 2.2 Requirements of a wavelet function $\psi \in L^2(\mathbb{R}^d)$ ($\hat{\psi}$ is the Fourier transform of the wavelet function ψ)

	Description	Condition
a	The average value of the wavelet in the time domain should be zero	$\int_{-\infty}^{\infty} \psi(t)dt = 0$
b	The function must have finite energy	$\int_{-\infty}^{\infty} \lvert\psi(t)\rvert^2 dt = 1$
c	Admissibility . The inverse wavelet transform only exists for $0 < C_\psi < \infty$. This means that the analyzed signal can be reconstructed without loss of information. The constant C_ψ is called the admissibility constant	$C_\psi = 2\pi \int_{-\infty}^{\infty} \frac{\lvert\hat{\psi}(\omega)\rvert^2}{\lvert\omega\rvert} d\omega$

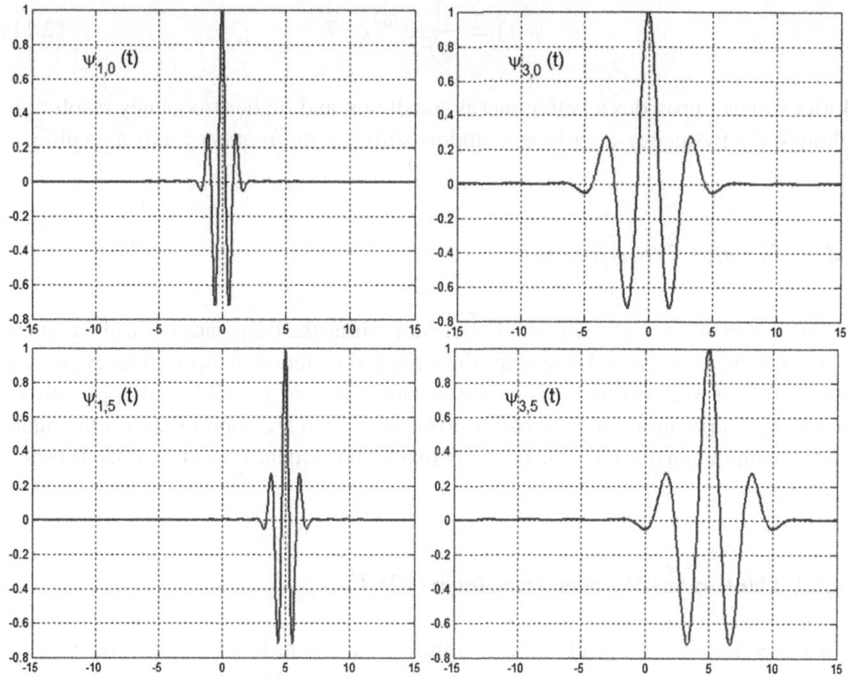

Fig. 2.1 Morlet's "wavelet daughters" $\psi_{s,\tau}(t)$, that are dilated and translated functions derived from the "mother wavelet" $\psi(t)$, according with Eq. 2.9

expressed as (Eq. 2.9):

$$\psi_{s,\tau}(t) = \frac{1}{\sqrt{s}}\psi(t)\left(\frac{t-\tau}{s}\right) \tag{2.9}$$

where s, τ are real and s>0. Wavelets expressed by (2.9) include an energy normalizatation $s^{-1/2}$ which keeps the energy of the daughter wavelets the same as the energy of their mother.

Morlet wavelet is the most popular complex wavelet used in practice, which mother wavelet is defined as Eq. 2.10 and represented in Fig. 2.1

$$\psi(t) = \frac{1}{\sqrt[4]{\pi}}\left(e^{j\omega t} - e^{-\frac{\omega^2}{2}}\right)e^{-\frac{t^2}{2}} \tag{2.10}$$

where ω is the central frequency of the mother wavelet. Note that the term $e^{-\frac{\omega^2}{2}}$ is used for correcting the non-zero mean of the complex sinusoid, and it can be negligible for $\omega < 5$

For practical purposes, for large ω, e.g., $\omega > 5$, Eq. 2.10 can be simplified [38] by taking a complex cos wave modulated by a Gaussian envelope (Eq. 2.11):

$$\psi(t) = \frac{1}{\sqrt[4]{\pi}} e^{j\omega t} e^{-\frac{t^2}{2}} \tag{2.11}$$

Morlet wavelet provides a better energy localizing and higher frequency resolution, although the frequency-coordinate window shifts along frequency axis as scaling.

2.4.2 *Wavelet Transforms*

The basic aim of wavelet analysis is both to determine the frequency (or scale) content of a signal and to assess and determine the temporal variation of this frequency content [39]. Although wavelet analysis covers a wide range of methods and applications, fundamental operations are *wavelet transforms*, which are appropriate for the many natural phenomena that have the property that high frequency events happen for short durations.

2.4.2.1 Continuous Wavelet Transform (CWT)

A wavelet transform correlates the signal with a family of waveforms $\psi_{s,\tau}$ or *wavelets*—it is also called time-frequency atoms by Mallat [40] and kernel by other authors—that meet the conditions of Table 2.2. The corresponding continuous wavelet time-frequency transform of $f \in L^2(\mathbb{R})$ is expressed by (Eq. 2.12):

$$\mathscr{W}_{\psi,s,\tau}\{f(t)\} = \int_{-\infty}^{\infty} f(t) \cdot \psi^*_{s,\tau}(t)dt = \langle f, \psi_{s,\tau} \rangle \tag{2.12}$$

where $\mathscr{W}_{\psi,s,\tau}\{f(t)\}$ is the CWT of f(t) with basis function family $\psi_{s,\tau}$.

The wavelet coefficients represent a measure of similarity in the frequency content between a signal and a chosen wavelet function. These coefficients are computed as a convolution of the signal and the scaled wavelet function, which can be interpreted as a dilated band-pass filter because of its band-pass like spectrum.

2.4.2.2 Continuous Wavelet Transform with Discrete Coefficients (CWTDC)

Signals are usually band-limited, which is equivalent to having finite energy, and therefore just a constrained interval of scales is useful. However, the continuous wavelet transform produces redundant information when capturing all the characteristics of the signal. The Discrete Wavelet Transform (CWTDC) has been created to minimize the redundancies produced by CWT.

It is possible to compute the wavelet transform for just a proper selection of values of the frequency and time parameters and still not loose any information as recovering the original time series from its transform. The continuous wavelet transform with discrete coefficients is very similar to the continuous wavelet transform, but the parameters s and τ are fixed to the power of 2, following these expressions:

$$\tau = 2^{-j}n,\ s = 2^{-j},\ \text{for m} \geq 0 \text{ and n} \in (-\infty, \infty) \tag{2.13}$$

The functional relationship between daughter $\psi_{k,j}(t)$ and mother $\psi(t)$, in the scale k and displacement j, is expressed as:

$$\psi_{k,j}(t) = 2^{j/2}\psi(2^j - k) \tag{2.14}$$

So, Eq. 2.14 is a special case of Eq. 2.9. From the continuous wavelet function (Eq. 2.12) and the new values of s and τ from Eq. 2.13, CWTDC takes the form of Eq. 2.15.

$$\mathscr{W}_{\psi,k,j}\{f(t)\} = \sqrt{2j} \int_{-\infty}^{\infty} f(t) \cdot \psi^*_{k,j}(2^j t - k)dt \tag{2.15}$$

2.4.2.3 Discrete Wavelet Transform (DWT)

Although the output of the continuous wavelet transform contains discrete coefficients, its implementation could be hard, since the input signal is continuous. Discrete wavelet transform is the alternative.

If f $\in L_2(\mathbb{R})$ and h $\in L_1(\mathbb{R})$, the convolution between the two signals: $g(x) = (f \otimes h)(x) = \int_{-\infty}^{\infty} f(t)h(x-t)dt$. The continuous wavelet transform of a f(t) signal, given at Eq. 2.12, yields a infinite set of wavelet coefficients. In a discrete form, where a time series u is a discrete sequence values of $(u_1, u_2, \ldots, u_n)$ separated in time by a constant time interval δt, the expression for the wavelet coefficient is given at time index j and scale s in the expression 2.16.

$$\mathscr{W}_{\psi,s,j}(u_n) = \sum_{n=0}^{N-1} u_n \psi^* \left[\frac{(n-j)\delta t}{s}\right] \tag{2.16}$$

where u_n is the discrete sequence, N denotes the length of the studied time series, ψ^* is the wavelet complex conjugated, and δt denotes the sampling period.

The algorithm to calculate the wavelet transform from Eq. 2.16, is a loop of convolutions performed N times for each scale, where N is the number of elements in the time series.

The convolution theorem states that the Fourier transform of the convolution of two functions is the product of Fourier transforms of each function $\mathscr{F}\{u(t) \otimes y(t)\} = \mathscr{F}\{u(t)\} \otimes \mathscr{F}\{y(t)\}$. Performing the Fourier transform in both sides of the Eq. 2.16, the inverse of wavelet transform can be expressed as Eq. 2.17. Applying the Fourier inverse transform to $\mathscr{F}\{\mathscr{W}_{\psi,s,j}(u_n)\}$, and following the work of Grinsted et al. [26], an efficient algorithm has been created to compute the discrete wavelet transform for the natural time series of this dissertation. This DWT can be computed by the inverse Fourier transform, as the Eq. 2.17 states.

$$\mathscr{F}\left\{\mathscr{W}_{\psi,s,j}(u_n)\right\} = \sum_{k=0}^{N-1} \hat{u}_k \cdot \hat{\psi}^*(s\omega_k)e^{i\omega_k n\delta t} \tag{2.17}$$

where k = 0 …N−1 is the frequency index, the Fourier transform of the function $\psi(s, t)$ is $\hat{\psi}(s\omega_k)$. The angular frequency ω_k admits the following values:

$$\omega_k = (2\pi k)/(N\delta t) \text{ for } k \leq N/2$$
$$\omega_k = -(2\pi k)/(N\delta t) \text{ for } k > N/2$$

2.4.2.4 Ouliers Detection in the Frequency Domain

Since outlier is an observation with characteristics of high-frequency phenomena, then wavelet technique is an excellent tool as outlier detector because of good location of frequencies. Most of energy and information of the data are usually concentrated in the first few coefficients. The outliers, as the noise, are on high frequency bands and reside in high-order coefficients. Therefore, the true data and outliers can be separated in the wavelet space (frequency domain).

In a discrete wavelet analysis, a signal u(t) can be represented by a decomposition of the signal into approximations A_j and detailed coefficients D_j. This is accomplished using shifted and scaled versions of the original (mother) wavelet as given in Eq. 2.16. In practice, the wavelet coefficients are computed efficiently using the *pyramid algorithm*, introduced in the context of multiresolution analysis by Mallat [40], that is based on a pair of high and low pass filters. The DWT in Eq. 2.16 produces a matrix of coefficients $[c_{i,k}]$. Then, the function can be represented by:

$$u(t) = \sum_{i=-\infty}^{j} \sum_{k} c_{i,k}\psi_{i,k}(t) = \sum_{i=-\infty}^{j-1} \sum_{k} D_{i,k}\psi_{i,k}(t) + \sum_{j} A_{j,k}\psi_{j,k}(t) \tag{2.18}$$

A_i is the approximation component set of the time series, and represents the low-frequency content. This low pass filter is like to continuously calculate a moving average of weighted data. D_i is the detailed component set of the time series, and represents the high-frequency content. This high pass filter consists on a moving difference of the data. These two sets of wavelet coefficients facilitate the recursive form of the *pyramid algorithm* [40].

Struzik and Siebes [41] propose a methodology capable of determining the statistical nature of the non-stationary process. The method checks the internal consistency of the scaling behavior of the process within the paradigm of the multifractal spectrum. Deviation from the expected spectrum is interpreted as the potential presence of outliers. Chinarro et al. [11] proposed a *wavelet-rosner test* and applied it to the time series of a karst aquifer. The method also stems from the wavelet based multiresolution analysis. After checking the A_i and D_i coefficients (Eq. 2.18) to detect outliers, Rosner's test is applied to remove or replace the abnormal values. A brief description of method in steps is:

1. Firstly, to avoid null values, a temporary shrunk series has been created with all elements from the raw data except null elements.
2. Transform the time series to the wavelet domain. DWT decomposes the time series and computes the approximation coefficients vector A_i and detailed coefficients vector D_i, at level 1, following (2.18).
3. Apply the Rosner test on the A_i coefficients to get outliers in the frequency domain. On this step, another method to remove outliers can be used, but Rosner's test has been weel tested in hydrological series.
4. Eliminates all outliers from the A_i and from analogous index in D_i, then two shrunk vector A_r and D_r are created.
5. To restore the time series, use A_r and D_r to compute the inverse wavelet transform.

2.4.3 Wavelet Power Spectrum (WPS)

The wavelet power spectrum helps to estimate the repartition of energy in the signal to determine the concentration of a signal in singular instants and frequencies. Temporal variation in the distribution of energy across scales is one of the most usual applications of the wavelet transform.

The Wiener-Khinchin theorem states that energy spectral density of a function is the Fourier transform of the corresponding autocorrelation sequence [28]. Analogously, wavelet power spectrum ($\mathscr{P}_{\psi,s,\tau}$) of signal u(t) is defined as autocorrelation function of the wavelet transformation ($\mathscr{W}_{\psi,s,\tau}$) of u(t), and describes the power of the signal u(t) at a certain time τ on a scale s:

$$\mathscr{P}_{\psi,s,\tau}\{u(t)\} = \mathscr{W}_{\psi,s,\tau}\{u(t)\} * \mathscr{W}^*_{\psi,s,\tau}\{u(t)\} = \left|\mathscr{W}_{\psi,s,\tau}\{u(t)\}\right|^2 \tag{2.19}$$

Because the wavelet function $\psi(\tau)$ is in general complex, the wavelet transform $\mathscr{W}_{\psi,s,\tau}$ is also complex, with a real part, $\Re(\mathscr{W}_{\psi,s,\tau})$, an imaginary part, $\Im(\mathscr{W}_{\psi,s,\tau})$, an amplitude $\left|\mathscr{W}_{\psi,s,\tau}\right|$, and a phase, $\Re(\mathscr{W}_{\psi,s,\tau})/\Im(\mathscr{W}_{\psi,s,\tau})$. Analogously, WPS can be expressed in the same components.

Wavelets add a new dimension in the spectral analysis to work simultaneously with time and frequency. Wavelet power spectrum is, in fact, a three-dimensional depiction, with time on the x-axis, frequency or scale on the y-axis, and the z-axis is to render the power magnitude at a particular time and frequency. This is a suitable tool for the spectral analysis of a non-linear system.

2.4.4 Wavelet Transformation Caveats

2.4.4.1 Cone of Influence (COI)

When applying the CWT to a finite length time series, the scalogram inevitably suffers from border distortions. The cause is that the values of the transform at the ends of the series cannot be accurately calculated because the transform calculus takes values outside the series range. These hedge effects also increase with s in a rate that depends on the mother function. The region in which the transform suffers from these edge effects is called the cone of influence (COI) and should be marked to take care in interpreting the belonged values [42]. One solution is to pad the end of the time series with zeroes before applying the wavelet transform and then remove them afterward. The padding is an extension of time series should be sufficient to spread out the time series to the next power of two. The zero padding reduces the variance, but introduces discontinuities at the endpoints and decreases the amplitude near the edges as going to larger scales [43].

As wavelet coefficients at COI suffer the same input discontinuity, the solution may be to rescale the remaining wavelet with choosing the *e-folding* time, i.e., as the distance at which the wavelet power drops by a factor e^{-2}. Larger e-folding time implies more expansion of the wavelet spectrum, The e-folding time is a measure of the wavelet width, relative to the wavelet scale s and ensures that the edge effects are negligible above a threshold for a given signal u(t) [26].

2.4.4.2 Choosing Wavelet Function

One singular characteristic of wavelet analysis is the arbitrary choice of the wavelet function. The below list is based on factors given by Torrence and Compo [43], in order to select a suitable wavelet to get best performance, and has been completed with other considerations by the author of the thesis:

(a) *Discrete or continuous.* DWT provides a more compact representation of data. So, DWT is rather suited for image processing, signal coding, noise reduction and computer vision. Nevertheless, CWT (also CWTDC) is a transformation that provides a high redundancy of data, and is suitable for time series analysis and feature extraction purposes.

(b) *Orthogonal or nonorthogonal.* The use of an orthogonal basis implies the use of DWT while a non-orthogonal wavelet function can be used with either the discrete or the continuous wavelet transform. Orthogonal wavelet functions have a zero correlation each other while non-orthogonal wavelets have a nonzero correlation. Using an orthogonal wavelet, the signal can be transformed to the frequency domain and then return to the time domain with a negligible loss of information. Orthogonal wavelet analysis is useful for signal processing because it gives the most compact representation of the signal. Non-orthogonal wavelets tend to surplus energy, because of overlapping, and require a normalization to optimize the acquisition of information. They are useful for time series analysis.

(c) *Complex or real.* A complex wavelet function will return information about both amplitude and phase and is better adapted for capturing oscillatory behavior. A real wavelet function returns only power, but is useful in location of peak frequency.

(d) *Width.* A wide wavelet function will give good frequency resolution and a loss of time resolution while a narrow wavelet function will yield good time resolution but a poor frequency resolutions.

(e) *Shape.* The wavelet function should reflect the type of features to be presented. For time series with sharp jumps or steps, a box-like function would be better such as Harr's wavelet proposed by Haar [23]. Nevertheless, for smoothly varying time series, the recommendable wavelet is a smooth function such as a damped cosine. If a wavelet power spectra has to be performed, then "the choice of wavelet function is not critical, and any function will give the same qualitative results as another" [43].

(f) *Choice of scales.* In orthogonal wavelet analysis, the set of scales s is limited [44]. In non-orthogonal wavelet analysis, an arbitrary set of scales can be built up for a more complete plotting.

To analyze natural systems, this dissertation has chosen the Morlet wavelet in most cases, because of five interesting properties:

- The peak frequency, the energy frequency and the central instantaneous frequency of the Morlet wavelet are all equal facilitating the conversion from scales to frequencies.
- Heisenberg Box area has a reduced size with this wavelet, i.e., the uncertainty reaches a minimum value. Then Morlet wavelet has an optimal joint time-frequency concentration.
- The time radius and the frequency radius are equal; therefore, this wavelet represents the best compromise between time and frequency concentration.
- Finally, Morlet is a wavelet transformation that yields complex coefficients, with information on both the amplitude and phase. This facilitates the study of gaps and delays between two time series.

2.4.4.3 Advantages of Wavelet Transform

Some features can be observed in the application of wavelet transform and according to Strang [45] and Perrier et al. [46].

Nonlinearity. Analysis with Fourier transform is not completely successful in all types of problems. Exceptions are nonlinear systems, with very brief signals or sudden changes, as the typical time series given in a karst system and glacier discharge; hence, the study of their behavior should be carried out with different tools.

Stationarity. Most traditional mathematical methods that examine periodicities in the frequency domain, such as Fourier analysis, have implicitly assumed that the underlying processes are stationary. The wavelet transform is suitable for the analysis of non-stationary signals because it provides a better time and frequency localization properties, expanding time series into time frequency space, such that the intermittent periodicities can better be localized [46].

Global properties. A Fourier transform hides information about time. It proclaims unequivocally how much of each frequency a signal contains, but is unknowable about when these frequencies were emitted. Therefore, in Fourier transform, any instant of a signal is similar to any other, even if the signal is as complex as public clap echoes in a theater, or changes as radically as the runoff through a river after a severe rainstorm. For an application, wavelets only capture the local time-dependent properties of data; whereas Fourier transforms, due to space-filling nature of the trigonometric functions, can only capture global properties [47].

Computational efficiency Using the *big* O notation [48], the computational complexity of the discrete Fourier transform is $O(n^2)$, where n is a number of time samples. Fourier transform is overtaken by the Fast Fourier transform (FFT) with a complexity O(nlogn) which takes less steps to solve an instance of the same problem. However, this is still under the complexity function O(kn) for discrete wavelet transform in decomposition and reconstruction processes (Fig. 2.2). The transform upshot with wavelets can be implemented in a computer by a quicker and more efficient algorithm.

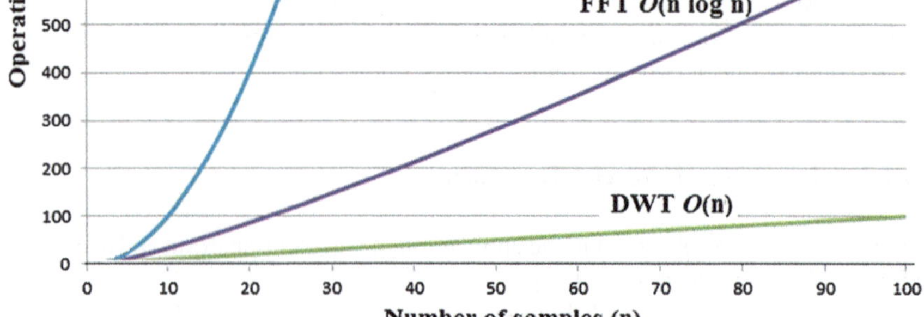

Fig. 2.2 Big-O complexity for Fourier Transform, FFT and DWT

2.5 Models

2.5.1 *Linear Time-invariant (LTI) models*

Most of the properties of linear time-invariant (LTI) systems are due to the fact that the system can be represented by linear differential (or difference) equations. Such properties include impulse response, convolution, duality, stability, scaling, etc. The properties of linear, time-invariant system should not in general apply to nonlinear systems. Nevertheless, LTI could be a first approach to identify the non-linear system.

The effect of any invariant linear system (LTI) on an arbitrary input signal is obtained by convolution of the input signal with the system's impulse response function. In a LTI system, the output of the system y(t) for an input x(t) can be obtained by the convolution integral:

$$y(t) = g(t) \star x(t) = \int_0^t g(t-\tau)x(\tau)d\tau \tag{2.20}$$

where g(t) is the *impulse response* of the system. That is, g(t) is the output of the system with an input x(t) = $\delta(t)$, where $\delta(t)$ is the Dirac delta. The impulse response completely characterizes the dynamic behavior of the system.

Applying the Laplace transform to the convolution integral (Eq. 2.20) we obtain Eq. 2.22 :

$$\mathscr{L}[y(t)] = \mathscr{L}[g(t) * x(t)] = \mathscr{L}[g(t)]\mathscr{L}[x(t)] \tag{2.21}$$

or in simple expression:

$$Y(s) = G(s)X(s) \tag{2.22}$$

where Y(s), G(s) and X(s) are the Laplace transforms of y(t), g(t) and x(t) respectively.

A **Transfer Function** (TF) is the mathematical representation of the relation between the input and output of a system. In a LTI system, TF can be expressed as the ratio of the Laplace transform of the output and the input, and corresponds to the Laplace transform of the impulse response G(s).

$$G(s) = \frac{Y(s)}{X(s)} \tag{2.23}$$

The transfer function of a system is rational fraction with numerator and denominator polynomials of the complex variable s:

$$G(s) = \frac{b_m s^m + \cdots + b_0}{a_n s^n + \cdots + a_0} = \frac{N(s)}{D(s)} \tag{2.24}$$

The roots of N(s) are called *zeros* of the system and roots of D(s) are called *poles* of the system. Poles and zeros are complex numbers that determine the dynamic behavior of the system. The real part of the poles defines the temporal prevalence of the term and the imaginary part its oscillatory behavior. The Transfer Function also characterizes the frequential behavior of a system, that is, how it responds to signals with different frequency components.

Most technological or natural systems are continuous and the signals are involved in dense time. However, when signals have to be processed by a computer in order to monitor, communicate or control a given system, they must be sampled at discrete points in time. The time interval between two sampling instants is the sampling interval or period (T). The models describing the behavior become discrete in time, and the discrete transfer function can be expressed by the *Z transform.*

$$\mathscr{Z}\{x[n]\} = \sum_{n=-\infty}^{\infty} x[n]z^{-n} \tag{2.25}$$

The concept of **Z transform** plays the same role for discrete time or sampled systems as the Laplace transform does for continuous time systems. Let $x^*(t) = \{x(kT)\}$ and $y^*(t) = \{y(kT)\}$ be the sequences obtained from the input and the output of the system at sampling instants ($t_k = kT$; $k = 0, 1, 2, \ldots$). The relationship between both data sequences is the convolution sum.

$$y^*(t) \sum_{k=0}^{\infty} x(kT)g_T^*(t-kT) = \sum_{k=0}^{\infty} g_T(kT)x^*(t-kT) = x^*(t) \star g_T^*(t) \tag{2.26}$$

where $g^*(t) = \{g_T(kT)\}$ is called the discrete impulse response of the system with sampling interval T. The sequence $g^*(t)$ is related with the impulse response of the continuous system g(t), but it is not the result of their sampling.

Linear systems are also modelled by discrete LTI Transfer Function, defined as the ratio of the *Z transform* of the output and the *Z transform* of the input. This function is rational with numerator and denominator polynomials of the complex variable z:

$$G_T(z) = \frac{Y(z)}{X(z)} = \frac{b_0 + \cdots + b_m z^{-m}}{a_0 + \cdots + a_n z^{-n}} \tag{2.27}$$

Thus, the activity of the system can be described as a set of parameters a_i, b_i of Eq. 2.27.

A useful format for the discrete LTI transfer function is to describe it in terms of z^{-1} because this is the unit delay operator. Roots of N(z) are called zeros and roots of D(z) are called poles of the system. As in time-continuous systems, the poles and zeros determine the dynamic behavior of the system.

2.5.2 Frequential Transfer Function Model

Let the input of a LTI system be of the form $x = x_0 cos(\omega t)$, it is well known that the response $y(t)$ of the system in steady state is also sinusoidal $y(t) = y_0 cos(\omega t + \phi)$. If $G(s)$ is the transfer function of the system then:

$$
\begin{aligned}
y_0 &= x_0 |G(j\omega)| \\
\phi &= arg[G(j\omega)]
\end{aligned}
$$

$G(j\omega)$, that is the transfer function evaluated in $j\omega$, is called the frequential transfer function of the system and the following holds:

$$
G(j\omega) = \frac{Y(\omega)}{X(\omega)} \tag{2.28}
$$

where $X(\omega)$ and $Y(\omega)$ are the Fourier transforms of input and output, respectively. The frequential transfer function characterizes the frequential behavior of the system, that is, how the frequential components of the input are modified (amplitude change and phase delay) to compose the output. There are some graphical representations of $G(j\omega)$ such as the Bode and Nyquist diagrams that allow a straightforward analysis of the system in function of frequency. Another expression of frequential components of a signal, similar to the Fourier transform, is the power density spectrum, $\phi_x(\omega)$, (signal energy/frequency unit). Equation 2.29 depicts the power density spectrum as the square of the absolute value of its Fourier transform.

$$
\phi_x(\omega) = |X(\omega)|^2 \tag{2.29}
$$

2.5.3 Nonlinear Models

Nonlinear system identification from input-output data can be performed using general types of nonlinear models such as neuro-fuzzy networks, neural networks, Volterra series or other various orthogonal series to describe nonlinear dynamics.

Non-linear relationships between input and output data provide much flexibility to describe a system. Models to identify nonlinear systems are discussed in Haber [49], and an extensive bibliography classified by nonlinear identification techniques can be found in Giannakis and Serpedin [50]. The use of functional analysis as a tool for the study of nonlinear systems was initially conceived by Wiener [51]. Following this work, Singleton [52] and Bose [53] made pioneering efforts toward engineering applications who left a firm foundation in the theory of functional analysis for both discrete and continuous systems. Other interesting works studied the theory of nonlinear continuous systems using power series Volterra functional [54] and orthogonal expansion of the functional-G Weiner [55]. Many efforts to develop analysis techniques are related to nonlinear systems, with significant number of people from the Russian school: Liapounoff, Andronov, Chaikin, Kryloff, Bogoliuboff, mentioned by Minorsky [56].

2.5.3.1 Volterra Series

The *Volterra series*, a discrete version of Kolmogorov-Gabor polynomial, was originally developed to describe the nonlinearity of a very general class of nonlinear time-invariant process. Although the Volterra series representation of nonlinearity provides theoretical understanding of nonlinearity, the number of coefficients in this model is excessive and needs enormous requirements on the identification procedure (quality and quantity of data). A large class of nonlinear functionals can be represented in the form of a Volterra series, which maps input signals u to output signals. A nonlinear system can be modeled as an infinite sum of multidimensional convolution integrals of increasing order (2.30).

$$y(t) = \sum_{i=1}^{n} \left| \int_{0}^{\infty} \right|_{i} h_i(\tau_1, \ldots, \tau_n) \prod_{i=1}^{n} [x(t - \tau_i) d\tau_i] \tag{2.30}$$

where h(τ) is the kernel of convolution. Identifying the nuclei of the Volterra series is the essential problem. In the case n = 1 (a unique term in the development), Volterra series is a first order and corresponds to linear modeling. So, Volterra theory is a generalization of the linear convolution integral approach often applied to linear, time-invariant systems. The behavior of the model depends on the kernels of the integral functionals, and it is these functions that are to be identified. A closely related model was introduced by Wiener, with the form:

$$y(t) = \sum_{n=0}^{\infty} [G_n(k_n, u)]\,(t) \tag{2.31}$$

where the functionals G_n are also integral equations in (2.30), with kernels $k_n = \tau_1, \ldots, \tau_n$ (these were used by Wiener as a sort of orthogonalized version of the Volterra kernels, via a Gram-Schmidt procedure). In discrete-time models, the integrals are replaced by sums.

Nonlinear Volterra theory is widely developed and applied in several fields of science and technology, as an approach to the modeling of nonlinear system behavior. Schetzen [57] goes into more depth about the theory of Volterra and Wiener functionals. Leontaritis and Billings [58] discuss available identification methods for the kernels.

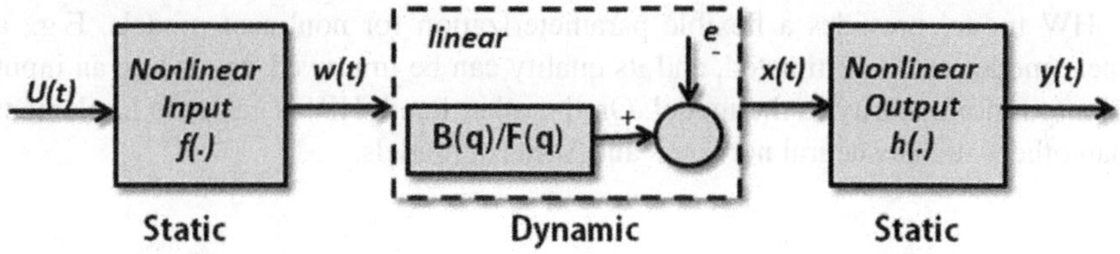

Fig. 2.3 Hammerstein-Wiener model

2.5.3.2 Hammerstein-Wiener Models

An alternative to drive problems of identifying a nonlinear model from input-output data is to use block-oriented nonlinear models consisting of static nonlinear function and linear dynamics subsystem such as Hammerstein model, Wiener model and feedback block-oriented model [59]. When the nonlinear function precedes the linear dynamic subsystem, it is called the Hammerstein model, whereas if it follows the linear dynamic subsystem, it is called the Wiener model.

Block-oriented models can be a flexible alternative in system identification tasks, due to they behold both linear and nonlinear features. Block-oriented models provide structures to study non-linear systems, under the hypothesis of LTI systems with static (no memory) non-linearity.

The model described by Eskinat et al. [60] is an interconnected model in cascade. Wiener model is similar [61], but the order of linear and non-linear blocks are inverted. The feedback block-oriented model consists of a static nonlinearity in the feedback loop of an LTI system [59].

In the Hammerstein-Wiener (HW) model (Fig. 2.3), a linear block model is the central block, represents the system dynamics and can be expressed by an output error polynomial model as Fig. 2.6. The first and last blocks are two nonlinear blocks w(t) = f(u(t)) and y(t) = h(x(t)). Where u(t) and y(t) are the inputs and outputs of the system and w(t) and x(t) are the input and output of the internal linear block. Hammerstein is a submodel with a nonlinear component followed by a linear component. Reversely, Wiener is a submodel with a linear component followed by a nonlinear component (Fig. 2.3). Nonlinear block models represent the static nonlinearities in the system [60] and can be identified following the parametrization given in (2.32).

$$\mathcal{N}(q) = \sum_{i=1}^{n} p_i g_i(x), \;\; p = [p_1, \ldots, p_n]^T \tag{2.32}$$

where g_i are a set of specified basis functions such as polynomial expressions, deadzone, saturation, piecewise, sigmoidnet or wavenet, look-up tables and fuzzy models [62]; and the p_i are the weights.

HW model provides a flexible parameterization for nonlinear models. E.g., a linear model can be estimated, and its quality can be improved, by adding an input or output nonlinearity to the model. On the other hand, HW is easier to implement than others such as neural networks and Volterra models.

2.6 System Identification Problem

The prior sections deal with preparation data (Sect. 2.2), time series analysis (Sect. 2.3), and review of model structures (Sect. 2.5), that are necessary in the identification process.

2.6.1 Posed Issue

The general problem of the system identification is to determine the most suitable model for the system, such as the problem graphically posed in the Fig. 2.4.

The model $g(u(t), \vartheta)$ maps the input u(t) to the output y(t) which is corrupted by perturbations $\varepsilon(t)$. The verification of the model consists of finding the vector ϑ that minimizes the error between real data and predicted data: e(t)= y(t)-$\hat{y}(t)$.

The searching of a model of a system from observed input-output data, implies the **input-output data** in the preparation stage, a set of candidate **model structures**, and some **criteria** to select a particular model in the set (Fig. 2.5).

System identification is carried out through stages of preparation, analysis and the identification as such (selection and optimization). The analysis procedures try to obtain the most possible details inside the system and take useful information from the time series. E.g., the elucidation if it is a linear system or a nonlinear system, a time-invariant system or a time-variant system a continuous system or a discontinuous system, a single input system or a multi input-multi output system, an open-loop system or a closed-loop system, etc (Fig. 2.5).

The selection stage is the identification of a suitable, identifiable model structure. The final identification stage—with the most critical procedures of

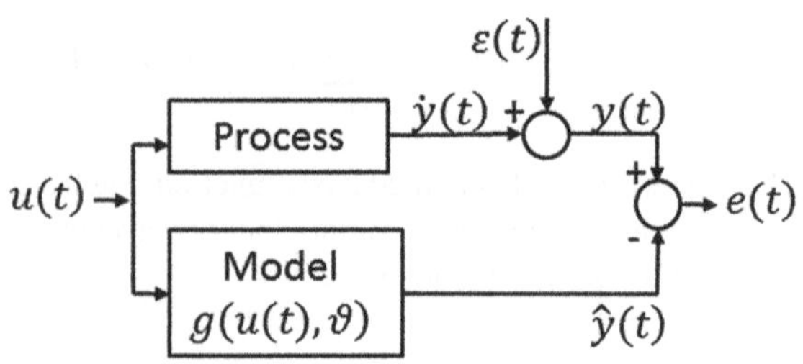

Fig. 2.4 Identification problem

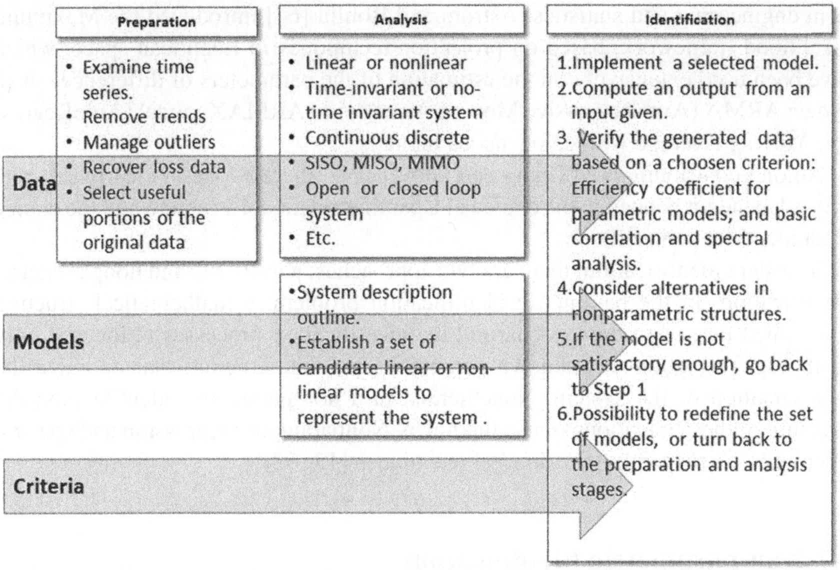

Fig. 2.5 Stages and components in the system identification problem

estimation—computes the best model according to the data and a given criterion. The validation of model measures the ability of the model to explain the observed data different to those used in the model identification and estimation stages, as the general prediction error (maximum likelihood) and efficiency criterion for parametric models; and basic correlation and spectral analysis in nonparametric structures [62]. As each system requires a different model to be chosen among those explained in Sect. 2.5, the process has to be necessary iterative and sometimes applying an *ad hoc* method.

2.6.2 The Literature Highlights

This section can not describe all historical processes on the system identification, only some highlights are mentioned from the engineering views. A further purpose can be found in the work of Deistler [63], with an excellent review of the history of system identification and time series analysis.

Spectrum analysis of time series might have commenced in 1664, when Isaac Newton decomposed a light signal into frequency components as passing the light through a glass prism. In 1800, Herschel measured the average energy in various frequency bands of the spectrum by placing thermometers along each band.

The first foundations on identification processes were set by mathematicians—Gauss (1809) and Fisher (1912)—although with subsequent important contributions from engineering and statistics. Aström and Bohlin [64] introduced the Maximum Likelihood framework, based on projection techniques in Euclidean space, which have been used extensively on the estimation of the parameters of difference—also known ARMA (AutoRegressive Moving Average) or ARMAX model (AutoRegressive Moving Average with eXogeneous inputs).

Among contributions of Ljung and Glover [18], there is one to clearly separate two independent concepts: the choice of a parametric model structure and the choice of an identification criterion.

In system identification, there are two approaches: parametric and nonparametric identification. In the parametric identification problem, a mathematical structure is assumed to govern the system, and the identification processes is focused only on the determination of unknown parameters for this structure that optimize the representation of the system. Nonetheless, in a non-parametric identification the structure of these equations is also unknown. Nonparametric regression and spectral techniques correspond to this kind of techniques [13, 62].

2.7 Non-parametric Identification

Nonparametric identification techniques provide a very effective and simple way of finding model structure in data sets without the imposition of a parametric one. Commonly, the initial process to carry out is the nonparametric identification, and then, if it were suitable, the parametric identification should be performed. The next sections review the non-parametric identification methods from time domain and frequency domain perspectives.

2.7.1 Non-parametric Identification in the Time Domain

2.7.1.1 Cross-Correlation

Cross-covariance is a non-parametric identification technique and is related with the impulse response g(t) (Eq. 2.20) of a system and thus with its behavior [62, 65]. Assuming that the signals have zero mean, if $y^\star(t)$ and $x^\star(t)$ are uncorrelated, the correlation between the input and the output is:

$$y^\star(t) = \sum_{k=0}^{\infty} g(kT)x^\star(t-kT) + v^\star(t) \tag{2.33}$$

where $v^\star(t)$ is the noise in the system. The signals involved can be regarded as the realization of stochastic processes. We can define the following coefficients and functions: If $v^\star(t)$ and $x^\star(t)$ are uncorrelated, the (cross) covariance function between the input and the output is:

$$R_{xy}(\tau) = \sum_{k=0}^{\infty} g(kT)R_x(\tau - kT) = g^{\star}(\tau)\star R_x^{\star}(\tau)$$

That is, the cross correlation is the convolution between the impulse response and the autocorrelation of the input. Thus, the impulse response can be estimated from the covariance (correlation if both signals have zero mean) if the input is a white noise. However, this is not a common case. For example, in hydrological systems, we have no control over the time series that always differs from the white noise. This problem is solved by the use of a whitening filter over the input and the output [62, 65].

The so-called cross correlation coefficient (Eq. 2.34) [65] has been widely used to characterize karst systems:

$$r_{xy}(k) = \frac{C_{xy}^N(k)}{\sqrt{\theta_x^2\theta_y^2}} \tag{2.34}$$

where $C_{xy}^N(k)$ is an estimate of $R_{xy}^N(k)$ for a finite number of samples. Supposing that the input is white noise, the impulsional response of the system can be estimated (Eq. 2.35) as:

$$g(k) = \frac{r_{xy}(k)\sqrt{\theta_y^2}}{\sqrt{\theta_x^2}} \tag{2.35}$$

g(k) is a good estimator only if the input behaves as white noise.

2.7.1.2 The Impulse Response

When the system has a finite impulse response, the non-parametric identification is performed by an intermediate approach between non-parametric and parametric identifications, and corresponds with a representation of the system by a FIR structure. Two methods have been proposed to calculate the coefficients of the impulse response:

Method 1 is based on the Wiener-Hopf summation equation [66, 67]. The estimator of cross-correlation between the input x(t) and output y(t) of the system for an infinite time series, is given by:

$$R_{xy}^N(i) = \sum_{k=0}^{M-1} g(k)R_x(i-k) \quad \text{with } i = 0 \ldots M-1 \tag{2.36}$$

Deploying the Eq. 2.36 in a matrix form [67]:

$$\begin{pmatrix} R_x(0) & R_x(-1) & \cdots & R_x(1-M) \\ R_x(1) & R_x(0) & R_x(-1) & \cdots \\ \cdots & R_x(1) & R_x(0) & R_x(-1) \\ R_x(M-1) & \cdots & R_x(1) & R_x(0) \end{pmatrix} \cdot \begin{pmatrix} g(0) \\ g(1) \\ \cdots \\ g(M-1) \end{pmatrix} = \begin{pmatrix} R_{xy}(0) \\ R_{xy}(1) \\ \cdots \\ R_{xy}(M-1) \end{pmatrix} \tag{2.37}$$

Thus, the impulse response can be calculated as:

$$\mathbf{R}_x \cdot \mathbf{g} = \mathbf{R}_{xy} \Leftrightarrow \mathbf{g} = \mathbf{R}_x^{-1} \cdot \mathbf{R}_{xy} \tag{2.38}$$

Method 2 is based on a minimization process. The most commonly used are linear programming [68, 69] and least squares [70–72]. To identify the impulse response is to minimize the prediction error. Considering a FIR model of length M:

$$Q^\star(t) = \sum_{k=0}^{M-1} g(kT)x^\star(t-kT) + \epsilon(t) = \hat{Q}^\star(t) + \epsilon(t) \tag{2.39}$$

The minimization of $\epsilon(t)$ can be done by several methods. In karst hydrology, least squares, linear programming over the slack variables have been used.

The minimization of prediction error of a FIR model is a case included in the so called parametric identification methods [62]. However, the parametric identification considers also several model structures with infinite impulse response and several models for error behavior.

2.7.2 Non-parametric Identification in Frequency Domain

2.7.2.1 Spectral Analysis

Frequency response can be derived from Fourier transform of the impulse response signal. It provides information about the gain and phase of the system for different input frequencies. The cross spectrum can be calculated in the form:

$$\phi_{xy}^N(n) = \sum_{k=-M+1}^{M-1} w(k) R_{xy}^N(k) e^{-jn\frac{2\pi}{N}k} \tag{2.40}$$

Let $g_T^\star(t)$ the impulse response of a sampled system. Then, as it has been stated, the output $y^\star(t)$ from an input $x^\star(t)$, follows:

$$y^\star(t) = \sum_{k=0}^{\infty} g_T(kT)x^\star(t-kT) + v^\star(t) \tag{2.41}$$

where $v^{\star}(t)$ is the noise in the system. If $v^{\star}(t)$ and $x^{\star}(t)$ are uncorrelated, the (cross) covariance function between the input and the output is given by Eq. 2.34. Supposing the sequences of finite length N, applying the Fourier transform and the equations Eq. 2.34 the result is:

$$\phi_{xy}^{N}(\omega) = G^{N}(j\omega)\phi_{x}^{N}(\omega) \tag{2.42}$$

So, the cross-spectrum is defined as the Fourier transform of the cross-covariance. Thus, an estimate of the frequential transfer function can be obtained as

$$G_{xy}^{N}(j\omega) = \frac{\phi_{xy}^{N}(\omega)}{\phi_{x}^{N}(\omega)} \tag{2.43}$$

However, the result is a plot which is not useful directly for simulation purposes [3].

Another useful function derived from the cross spectrum is the Coherence Function, which is calculated by the following expression:

$$\gamma_{xy}^{2}(\omega) = \frac{|\phi_{xy}^{N}(\omega)|^{2}}{\phi_{x}^{N}(\omega)\phi_{y}^{N}(\omega)} \tag{2.44}$$

It measures the linear correlation between the input and the output of the system at each frequency ω. Notice also that the Coherence Function is dimensionless and can be shown that $0 \leq \gamma_{xy}^{2}(\omega) \leq 1$.

$G_{xy}^{N}(j\omega)$ characterizes the frequential behavior and the coherence $\gamma_{xy}^{2}(\omega)$ characterizes the linearity of a system.

2.7.2.2 The Cross Wavelet Spectrum (XWS)

The bivariate extension of wavelet analysis is recommended when the system involves two time series, instead of only one, to assess time-varying spectral relations between two signals which are often non-stationary.

Cross-wavelet transform (XWS)—also called *wavelet cross-scalogram* or *coscalogram*—gives information on the dependence between two signals, u(t_i) and y(t_i), as a function of time, similar to *cross-correlogram* (or the *cross-spectrogram*). XWS, a bivariate extension of WPS, is commonly applied in earth sciences, e.g., in the experimental cases of this document, the air temperature and glacier discharge or rainfall and conductivity in an aquifer system. The cross wavelet spectrum, introduced by Hudgins et al. [73] to study the atmosphere turbulence, reveals how regions in the time frequency space with large common power have a consistent relationship. This fact suggests a causality between both time series, and the expectation value of the correlation of two signals u(t) and y(t) is:

$$\mathscr{X}_{\psi,s,\tau}\left\{u(t), y(t)\right\} = \mathscr{W}_{\psi,s,\tau}^{*}\left\{u(t)\right\} * \mathscr{W}_{\psi,s,\tau}\left\{y(t)\right\} \tag{2.45}$$

Since the wavelet function $\psi(\tau)$ is in general complex, $\mathscr{X}_{\psi,s,\tau}\{u(t), y(t)\}$ is also complex.

The cross wavelet power of two signals describes the covariance between these times series at each scale of frequency. Cross wavelet spectrum illustrates quantitatively the similarity of power between two times. It has already been applied to rainfall-runoff cross analysis by Labat et al. [74], and was briefly discussed by Maraun and Kurths [75].

If complex wavelets are used, such as the Morlet wavelet, its squared absolute value $|\mathscr{X}_{\psi,s,\tau}\{u(t), y(t)\}|^2$ or simply its absolute value $|\mathscr{X}_{\psi,s,\tau}\{u(t), y(t)\}|$ is often gotten for better plotting. The value of $|\mathscr{X}_{\psi,s,\tau}\{u(t), y(t)\}|^2$ is large when $|\mathscr{P}_{\psi,s,\tau}\{u(t)\}|$ and $|\mathscr{P}_{\psi,s,\tau}\{y(t)\}|$ are big at the closeness of scales (frequencies) and around the same time, regardless of the local phase difference. When the phase information is required, the Eq. (2.45) should be expressed in terms of its module and phase angle:

$$\begin{aligned}\mathscr{X}_{\psi,s,\tau}\{u(t), y(t)\} &= |u(s,\tau)|\, e^{-i\theta_u(s,\tau)}\, |y(s,\tau)|\, e^{i\theta_y(s,\tau)} \\ &= \left|\mathscr{X}_{\psi,s,\tau}\{u(t), y(t)\}\right| e^{i\theta_y(s,\tau)-i\theta_u(s,\tau)} \end{aligned} \tag{2.46}$$

This means that the phase angle $i\theta_y(s,\tau) - i\theta_u(s,\tau)$ reflects the phase difference by which y(t) leads u(t) at the given scale and time. Van Milligen et al. [76] introduced *delayed wavelet cross spectrum*, a useful quantity to detect structures from two separated observation points.

In hydrology, wavelet cross-correlation should sometimes be preferred to classical cross-correlation, because this method provides new insights into the scale depending on the degree of correlation between two given signals [77, 78].

2.7.2.3 Wavelet Coherence Spectrum (CWS)

An extension to Fourier analysis, to allow for non-stationarity, is windowed Fourier analysis, but this overcomes the assumption of global stationarity within each segment. In the time-scale domain, cross-spectrum cannot be normalized locally assuming stationarity to have a value bounded by, for example, zero and one, because multiple points should be involved for some degree of smoothing.

Coherence in signal processing consists of a measure of the correlation between two signals. Power Spectrum represents the power carried by each frequency in a signal. The similarity of two signals can be checked by estimating CWS. Two examples coherence application are: the *wavelet local correlation coefficients* introduced by Buresti and Lombardi [79] to measure the phase coherence of the signals; and *cross wavelet coherence function* introduced by Sello and Bellazzini [80] to assess the intensity coherence of turbulent signals. In practice, to calculate the coherence of two signals, we calculate the cross-power spectrum, because the coherence is the normalized measurement of the cross-power spectrum and can be calculated dividing cross-power spectrum by the squared root of the product of the spectra of the signals

(Eq. 2.47). The expression of wavelet coherence is [81]:

$$\mathscr{C}_{\psi,s,\tau}\{u(t), y(t)\} = \frac{\left\langle \left| \mathscr{X}_{\psi,s,\tau}\{u(t), y(t)\} \right| \right\rangle}{\left(\left\langle \mathscr{P}_{\psi,s,\tau}\{u(t)\} \right\rangle \cdot \left\langle \mathscr{P}_{\psi,s,\tau}\{y(t)\} \right\rangle \right)^{1/2}} \tag{2.47}$$

The $\langle . \rangle$ operator was suggested by Torrence and Compo [43] as the best compromise solution, providing the minimum amount of smoothing necessary to include two independent points in time and scale. Polar expression is a useful form of wavelet coherence with modulus $\rho_{\psi,s,\tau}$ and phase $\phi_{\psi,s,\tau}$, such as:

$$\begin{aligned} \mathscr{C}_{\psi,s,\tau}\{u(t), y(t)\} &= \rho_{\psi,s,\tau}\{u(t), y(t)\} \cdot e^{i\phi_{\psi,s,\tau}(u(t),y(t))} \\ \phi_{\psi,s,\tau} &= \tan^{-1}\left[\mathrm{Im}\left(\mathscr{C}_{\psi,s,\tau}\right)/\mathrm{Re}\left(\mathscr{C}_{\psi,s,\tau}\right)\right] \end{aligned} \tag{2.48}$$

A value of 1 means a perfect coupling between u(t) and y(t) around time τ on a scale s for the wavelet ψ. For a zero or negative value the variations of two signals are not correlated, and for positive value between zero and one the variations are correlated in a certain degree. This value meaning is similar to a traditional correlation coefficient as defined by Barret [82], and it is useful to think of the wavelet coherence as a localized correlation coefficient in time frequency space. Due to these properties, wavelet coherence is an increasingly popular method in analyzing hydrological correlated events.

The fact a signal is correlated with other, not only means that some energy in a frequency is present in both signals, but that plus the frequency which is present in both signals is also related by phase.

The magnitude squared coherence is also used, which is the squared value of the cross-power spectrum divided by the product of the power of the spectra of both signals. The measure of coupling average in the scale (frequency) domain between input and output signals u(t) and y(t) would provide information about the relationship of the signals. This can be got by the wavelet cross spectrum coefficient or simply wavelet coherence spectrum (CWS), obtained from the normalized wavelet cross-spectrum (to have values between zero and one).

2.8 Parametric Identification

Parametric identification relies on a model previously defined by a set of parameters that must be calculated to accomplish a given quality criteria. E.g., the system characteristics can have a parametric representation through a polynomial of a finite and known degree. The model structure can be obtained by physical modeling (grey box) or it may be a standard one (black box). In the latter case, a set of generic standard structures must be taken into account (OE, FIR, ARX, ARMAX and BJ).

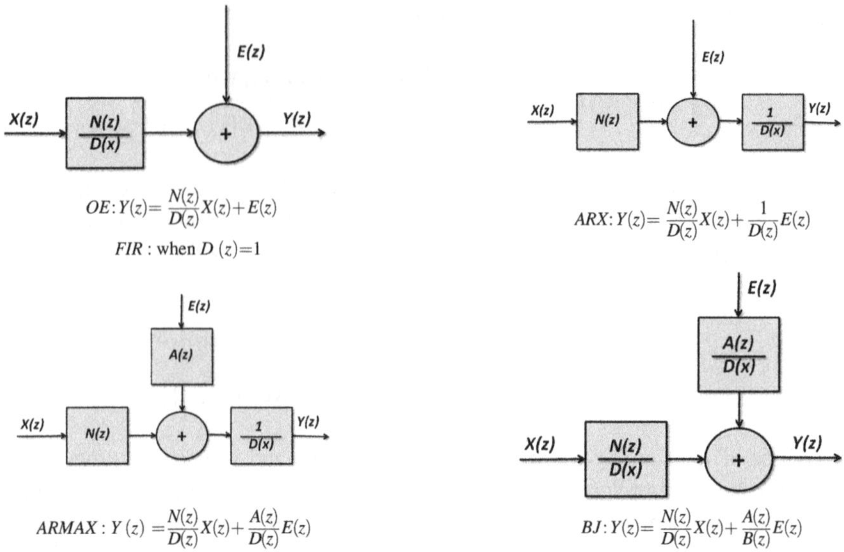

Fig. 2.6 System model structures

2.8.1 Linear Parametric Identification

Parametric identification techniques depend mostly on Prediction-Error Methods (PEM) [62]. The output of system $y^{\star}(t)$ can be expressed as Eq. 2.33. A more useful expression is based on the Z transform:

$$Y(z) = G(z)X(z) + W(z) \tag{2.49}$$

This expression can be rewritten as:

$$Y(z) = G(z)X(z) + H(z)E(z) = \frac{N(z)}{D(z)}X(z) + \frac{A(z)}{B(z)}E(z) \tag{2.50}$$

where $E(z)$ is the transform of a white noise, $\epsilon(t)$. $G(z)$, the transfer function of the system, and $H(z)$, the stochastic behavior of noise. $G(z)$ and $H(z)$ are rational functions whose numerator and denominator are polynomials of the z variable. The relationship between both functions defines several model structures. Figure 2.6 shows the most common ones: AutoRegressive eXogeneous (ARX) model, AutoRegressive Moving Average eXogeneous (ARMAX) model, Box-Jenkins (BJ) model and Output Error (OE) models. The features and advantages of each structure have been studied in several works, see for example Ljung [62].

The ARX model D(z)Y(z) = N(z)X(z) + E(z) is the easiest to estimate since the corresponding estimation problem is of a linear regression type. The foremost disadvantage is that the disturbance model 1/N(z) comes along with the system's poles. It is, therefore, easy to get an incorrect estimate of the system dynamics because the A (Eq. 2.50) polynomial can also include the disturbance properties. So, higher orders in A and B coefficients in the Eq. 2.50 may be required. If the signal-to-noise ratio is good, this disadvantage is less important.

The ARMAX model D(z)Y(z) = N(z)X(z) + A(z)E(z) has more flexibility in the handling of disturbance modeling than the ARX model. For this reason, ARMAX is a widespread used model and performs well in many engineering applications.

The OE model Y(z) = [N(z)/D(z)]u(z) + E(z) has the advantage that the system dynamics can be described separately and that no parameters are wasted on a disturbance model. If the system operates without feedback during the data collecting, a correct description of the transfer function G(z) = N(z)/D(z) can be obtained regardless of the nature of the disturbance.

In the BJ model Y(z) = [N(z)/D(z)]u(z) + [A(z)/B(z)]E(z) the disturbance properties are modeled separately from the system dynamics.

2.8.2 Selection and Verification Criteria

To get a model reliable, the results and predictions inferred from model should be verified and validated. Model validation is carried out by comparing the model behavior with the system's one and evaluating the difference. All models have a certain domain of validity. This may determine how exactly they are able to describe the system behavior.

Efficiency criteria can be defined as mathematical measures of how well a model simulation fits the available observations [83]. A number of different methods to set a criterion have been suggested in the literature, e.g., least squares, generalized least squares, maximum likelihood or instrumental variables. Krause et al. [84] have studied the utility of several efficiency criteria in three examples using a simple observed streamflow hydrograph, declaring that: "The selection and use of a specific efficiency criteria and the interpretation of the results can be a challenge for even the most experienced hydrologist, since each criterion may place different emphasis on different types of simulated and observed behaviors".

Nash-Sutcliffe efficiency, coefficient of determination, and index of agreement, are frequently applied to verify hydrologic models. The efficiency value E proposed by Nash and Sutcliffe [85] is defined as one minus the sum of the absolute squared differences between the predicted and observed values normalized by the variance of the observed values during the period (Eq. 2.51).

$$E = 1 - \frac{\sum_{k=1}^{N}\left[y(kT) - \hat{y}(kt)\right]^2}{\sum_{k=1}^{N}\left[y(kT) - \overline{y}\right]^2} \tag{2.51}$$

where $y(kT)$ is observed data in the sampling interval T (with k = 0, 1, 2, ..., N), $\hat{y}(kt)$ is modeled output, and $\overline{y}$ is mean of observed data. Nash-Sutcliffe efficiencies can range from $-\infty$ to 1. An efficiency value of 1 (E = 1) corresponds to a perfect match of model output to the measured data. An efficiency value of 0 (E = 0) indicates that the model is as accurate as the mean of the observed data, whereas an efficiency less than zero (E < 0) occurs when the observed mean is a better predictor than the model or, in other words, when the residual variance (described by the numerator in the expression above), is larger than the data variance (described by the denominator). Essentially, the closer the model efficiency value is to 1, the more accurate the model is.

According to Legates and McCabe [86], the disadvantage of the Nash-Sutcliffe efficiency is the fact that the differences between the observed and predicted values are calculated as squared values. As a result, larger values in a time series are strongly overestimated, whereas lower values are neglected. E.g., in the quantification of an aquifer discharge, this criterion could lead to an overestimation of the model performance during peak flows and an underestimation during low flow conditions. Nevertheless, Nash-Sutcliffe efficiency, expressed frequently as 0–1 coefficient (E), is the unique criterion used in the experimental cases of this thesis, in order to standardize the comparison between models.

2.9 Nonlinear Identification

It is difficult to establish a clear identification methodology of nonlinear systems, since analysis is usually more intricate than in the identification of linear models, because of the variety of nonlinear model structures and nonlinear behaviors.

Donoho and Johnstone [87], and Donoho [88] introduced nonlinear wavelet estimators in nonparametric regression through thresholding, i.e., the term-by-term assessment of coefficients in the wavelet expansion. Only coefficients that exceed a predetermined threshold are taken in account. This produces the wavelet shrinkage. Bendat [89] describes procedures to identify and analyze the properties of many types of nonlinear systems as Zero-Memory Nonlinear Systems and Parallel Nonlinear System, with analysis of Nonlinear System Input/Output Relationships. Zhang [90] applied wavelet theory for nonlinear system identification, with a wavelet basis as a universal function approximator, with a neural network used to determine the resolution, and the translation coefficients of the wavelet. This nonparametric estimator named wavelet network has a neural network like structure that makes use of techniques of regressor selection completed with backpropagation procedure.

This section is going to focus only on nonlinear parametric identification, and, inside this type, Volterra series and Hammerstein- Wiener methods.

2.9.1 Nonlinear Parametric Identification

2.9.1.1 Volterra Identification

Volterra series have been widely applied as nonlinear system modeling technique. These Volterra representations lead to very complicated identification algorithms since evaluation of higher order Volterra kernel often require very large amounts of data for even low order nonlinearities. However, kernels can be calculated for systems whose order is known and finite.

Some identification procedures to calculate higher-order kernels have given by [91], but the proposed solutions reduce their practical application only to model system of low non-linearity. Mirri et al. [92] suggested three different models for Volterra series, in order to reduce the number of kernels that are considered in the identification process of a non-linear system. The reduced number of mathematical operators involved, simplifies the experimental procedures, but it needs a somehow information about the system behavior.

2.9.1.2 Hammerstein-Wiener Identification

The literature about the Hammerstein-Wiener model identification is ample [93, 94]. Most important methods try to reduce the parameter redundancy by using linear and nonlinear structures. Other methods use optimized algorithms to decrease the computation complexity.

- The iterative algorithm, a classical method proposed by Narendra and Gallman [95], parametrizes the system for the linear prediction error. In the most common version of this method, the parameter set is usually divided into two subsets. One set is fixed while the other searches the optimal values. Then, both sets are switched to perform the reverse operation. The estimation is carried out by minimizing alternatively with respect to each set of parameters, a quadratic criterion on the prediction errors. A main problem of the iterative algorithm is the convergence.
- Bai et al. [96] studied the convergence properties of iterative algorithm from Narendra and Gallman [95] They show that the iterative algorithm with normalization is convergent in general, and for finite-impulse response, the convergence is reached in one step.
- A noniterative method was proposed by Chang and Luus [97] and show through numerical examples that the computation time by this method is considerably less than by the iterative, while the accuracy of the estimates is comparable.
- Bai [98] presented an optimal two stage identification algorithm for Hammerstein-Wiener model. The first step is the recursive least squares. The second one is the singular value decomposition of two matrices with fixed dimensions and do not increase as the number of the data point increases.
- Reference [99] described a blind approach, where the linear part is only using the output measurements, i.e., no information on the input.

- Geothals et al. [100] extend the unifying theorem for three subspace system identification algorithms by using Least Squares Support Vector Machines component to identify Hammerstein-Wiener structure.
- Wills et al. [94] illustrates a new maximum-likelihood based method for the identification of Hammerstein-Wiener model structures.

2.10 Conclusion

To study system dynamics, some stages are recommended: Preparation, analysis, model structures preselection and identification.

Raw data need refining and adjustment before time series analysis because of outliers or abnormal values in the readings. Among other criteria, Wavelet-Rosner is an efficient test for outlier detection in the frequency domain. Also, the selection of the proper sampling frequency is discussed.

In analysis under the scenario of the time domain, autocorrelation is a lag correlation of a given time series within itself, lagged by a number of times units. Positive autocorrelation is a specific form of 'persistence' of events, and time series is more predictable. The wavelet transform calculates the correlation between the signal and a wavelet function $\psi(t)$, the mother wavelet, parametrized by location and scale. The similarity between the signal and the wavelet function is computed separately for different time intervals, resulting in a three dimensional representation. In the frequency domain analysis, the power density spectrum assesses the energy level of the signal in different frequency bands.

Linear time-invariant models provide structures for systems. Although in nature all systems are nonlinear, in practical cases the use of linear models is justified in order to simplify the study or as a reference for other more exhaustive models. The effect of any invariant linear system (LTI) on an arbitrary input signal is obtained by convolution of the input signal with the system's impulse response function. A Transfer Function in LTI can be expressed as the ratio of the Laplace transform of the output and the input, and corresponds to the Laplace transform of the impulse response. The frequential transfer function characterizes the frequential behavior of the system, that is, how the frequential components of the input are modified (amplitude change and phase delay) to compose the output.

The non-linear models by Volterra equations are not suitable in the case of long uninterrupted records at a finer sampling rate, because Volterra models have a stationary or time invariant kernel. Nevertheless, block-oriented models have emerged as an appealing proposal due to their simplicity and their property of being valid over a larger operating region than a LTI model and are easier to implement than heavy-duty nonlinear models, such as neural networks and Volterra models.

Nonparametric identification techniques provide a very effective and simple way of finding model structure in data sets without the imposition of setting required parameters. Cross correlation is the convolution between the impulse response and the autocorrelation of the input. Thus, the impulse response can be estimated from the covariance if the input is a white noise. When the system has a finite impulse response, the non-parametric identification is performed by an intermediate approach between non-parametric and parametric identifications, and corresponds with a representation of the system by a FIR structure. A nonparametric identification in the frequency domain can be carried out by cross wavelet spectrum. This reveals how regions in the time frequency space with large common power spectrum suggest a causality between both time series. The coherence is the normalized version of cross wavelet spectrum.

Characterizing all the input-output properties of a system through exhaustive measurements is usually impossible. Instead, parametric identification is a way to make that a finite number of measurements help infer the system response to specified inputs. Parametric identification techniques, which depend mostly on Prediction-Error Methods (t), can be expressed by Z transform. Model validation is carried out by comparing the data generated by the model with those observed in the system. Nash-Sutcliffe efficiency, coefficient of determination, and index of agreement, are frequently applied to verify hydrologic models.

It is difficult to establish a clear identification methodology of nonlinear system identification, because of the variety of nonlinear model structures and nonlinear behaviors. These Volterra representations lead to very complicated identification algorithms since evaluation of higher order Volterra kernels often requires very large amounts of data for even low order nonlinearities. However, kernels can be calculated for systems whose order is known and finite. Most important methods in Hammerstein-Wiener model identification try to reduce the parameter redundancy by using linear and nonlinear structures. Other methods use optimized algorithms to decrease the computation complexity.

References

1. Williams T (1988) Modelling complex projects. Wiley, New Jersey
2. Eykhoff P (1974) System identification. Wiley, London
3. Ljung L, Glad T (1994) Modeling of dynamic systems. Prentice Hall, Englewood Cliffs
4. Alippi C, Anastasi G, Di-Francesco M, Roveri M (2009) Energy management in wireless sensor networks with energy-hungry sensors. IEEE Instrum Meas Mag 12(2):16–23
5. Hernàindez L, Baladrón C, Aguiar JM, Calavia L, Carro B, Sàinchez-Esguevillas A, Cook DJ, Chinarro D, Gómez J (2012) A study of the relationship between weather variables and electric power demand inside a smart grid/ smart world framework. Sensors 12:11571–11591
6. Zayed AI (1993) Advances in shannon's sampling theory [Hardcover]. CRC Press, Library of Congress, USA
7. Hawkins D (1980) Identification of outliers. Chapman and Hall, London
8. Barnett V, Lewis T (1994) Outliers in statistical data. Wiley, Chichester
9. Tabachnick B, Fidell L (1996) Using multivariate statistics. Harper-Collins College Publishers, New York
10. Rosner B (1983) Percentage points for a generalized esd many-outlier procedure. Technometrics 25:165–172
11. Chinarro D, Villarroel J, Cuchí J (2011) Wavelet analysis of fuenmayor karst spring, San Julián de Banzo, Huesca, Spain. Environ Earth Sci Spec Issue 65(8):1–13

12. Tukey J (1960) A survey of sampling from contaminated distributions. In: Olkin I et al (eds) contributions to probability and statistics: essays in honor of harold hotelling. Stanford University Press, Stanford, pp 448–485
13. Box G, Jenkins G (1976) Time series analysis: forecasting and control. Holden-Day, San Francisco
14. Stokes G (1879) Note on searching for hidden periodicities. Proc R Soc 29:122–125
15. Schuster A (1898) On the investigation of hidden periodicities with application to a supposed 26 day period of meteorological phenomena. Terr Magn Atmos Electricity (J Geophys Res) 3(1):13–41
16. Wellstead P (1981) Non-parametric methods of system identification. Automatica 17(1): 55–69
17. Welch P (1967) The use of the fast fourier transform for the estimation of the power spectra: a methods based on time averaging over short modified periopdograms. IEEE Trans Audio Electroacustic 15(2):70–73
18. Ljung L, Glover K (1981) Frequency domain versus time domain methods in system identification. Automatica 17(1):71–86
19. Blackman R, Tukey J (1958) The measurement of power spectra. Dover Publications, Inc., New York
20. Grossman A, Morlet J (1984) Decomposition of hardy functions into square integrable wavelets of constant shape. J Math Anal 15:732–736
21. Gabor D (1946) Theory of communication. J IEE 93(26):429–457
22. Kronland-Martinet R, Morlet J, Grossmann A (1987) Analysis of sound patterns through wavelet transforms. J Pattern Recognit 1(2):273–302
23. Haar A (1910) Zur theorie der orthogonalen funktionensysteme. Math Ann 69:331–371
24. Esteban D, Galand C (1977) Application of quadrature mirror filters to split-band voice coding schemes. In: Acoustics, Speech, and Signal Processing, IEEE International Conference on ICASSP-77, vol 2. pp. 191–195
25. Crochiere RE, Webber SA, Flanagan JL (1976) Digital coding of speech in subbands. Bell Syst Tech 85:1069–1085
26. Grinsted A, Moore JC, Jevrejeva S (2004) Application of the cross wavelet transform and wavelet coherence to geophysical time series. Nonlinear Proc Geophys 11:561–566
27. Sangbae Kima FI (2007) On the relationship between changes in stock prices and bond yields in the G7 countries: wavelet analysis. J Int Finan Markets, Inst Money 17:167–179
28. Ricker DW (2003) Echo signal processing (The springer international series in engineering and computer science). Kindle Edition
29. Rivera D, Lillo M, Arumí J (2007) ENSO influence on precipitation of chillan, chile: an approach using wavelets. Gestión Ambiental 13:33–48
30. Meyers SD, Kelly BG, O'Brien JJ (1992) An introduction to wavelet analysis in oceanography and meteorology: with application to the dispersion of yanai waves. Mon Weather Rev 121(13–31):101–105
31. Cuchí J, Villarroel J (2002) Análisis de respuesta de Fuenmayor (San Julián de Banzo, Huesca). Primeros resultados. Geogaceta 31:75–78
32. Labat D, Ababou R, Mangin A (1999a) Analyse en ondelettes en hydrologie karstique. Geoscience 329:881–887
33. Hernández L, Baladrón C, Aguiar JM, Carro B, Sánchez-Esguevillas A, Lloret J, Chinarro D, Gómez JJ, Cook D (2013) A multi-agent-system architecture for smart grid management and forecasting of energy demand in virtual power plants. IEEE Commun 51(1):106–114
34. Ebadi L, Shafri HZM, Mansor SB, Ashurov R (2013) A review of applying second-generation wavelets for noise removal from remote sensing data. Environ Earth Sci 70(6):2679–2690. doi:10.1007/s12665-013-2325-z
35. Amat S, Muñoz J (2008) Some wavelets tools for maxwell's equations. Carthagonova J Math 1(1):1–18
36. Daubechies I, Daniel Kleppner SM, Meyer Y, Ruskai MB, Weiss G (2001) Wavelets: seeing the forest and the trees. Office on Public Understanding of Science,Technical report, National Academy of Sciences

37. Daubechies I (1990) The wavelet transform time-frequency localization and signal analysis. IEEE Trans Inf Theory 36(5):961–1004
38. Foufoula-Gergiou E, Kumar P (1994) Wavelets in geophysics, (wavelet analysis and its applications), 1st edn. Academic Press, Boston
39. Heil CE (1989) Continuous and discrete wavelet transforms. SIAM Rev 31(4):628–666
40. Mallat S (1998) A wavelet tour of signal processing, 2nd edn. Academic Press, École Polytechnique, Paris, Courant Institute, New York University
41. Struzik Z, Siebes A (2000) Outlier detection and localisation with wavelet based multifractal formalism. Inf Syst (INS), R0008
42. Mallat S, Zhong S (1992) Characterization of signals from multiscale edges. IEEE Trans Pattern Anal Mach Intell 14(7):710–732
43. Torrence C, Compo GP (1998) A practical guide to wavelet analysis. Bulletin Am Meteorol Soc 79:61–78
44. Farge (1992) Wavelet transforms and their applications to turbulence. Annu Rev Fluid Mech 24:395–457
45. Strang G (1993) Wavelet transforms versus fourier transforms. Bulletin, new series, Am Math Soc 28:288–305
46. Perrier V, Philipovitch T, Basdevant C (1995) Wavelet spectra compared to fourier spectra. J Math Phys 36:1506–1519
47. Popivanov I, Miller RJ (2001) Similarity search over time series data using wavelets. Technical report TR CSRG 438. Department of Computer Science, University of Toronto, 438
48. Becher V, Bienvenu L, Downey R, Mayordomo E (2012) Computability, complexity and randomness (dagstuhl seminar 12021). Dagstuhl Rep 2(1):19–38
49. Haber R (1985) Nonlinearity test for dynamic processes. IFAC Identification and system parameter estimation symposium. Pergamon Press, Elmsford, New York, pp 409–414
50. Giannakis GB, Serpedin E (2001) A bibliography of non linear system identification. Signal Process 81:553–580
51. Wiener N (1958) Nonlinear problems in random theory. MIT Press, Cambridge
52. Singleton HE (1950) Theory of nonlinear transducers. Laboratory of Electronics, Massachusetts Institute of Technology, Cambridge, Technical Report 160
53. Bose AG (1959) A theory of nonlinear systems. Technical report, Research Laboratory of Electronics, Massachusetts Institute of Technology, Cambridge
54. Brilliant MB (1958) Theory of the analysis of nonlinear systems. Technical report, 345. Research Laboratory of Electronics, Massachusetts Institute of Technology, Cambridge
55. George DA (1959) Continuous nonlinear systems. Technical report, Research Laboratory of Electronics, Massachusetts Institute of Technology, Cambridge
56. Minorsky N (1947) Introduction to nonlinear mechanics. J. W. Edwards Press, Ann Arbor, Michigan
57. Schetzen M (1980) The volterra and wiener theories of nonlinear systems. Wiley, New York
58. Leontaritis I, Billings S (1985) Input-output parametric models for nonlinear systems. part i-deterministic non-linear systems. Int J Control 41:303–328
59. Pottmann M, Pearson RK (1998) Block-oriented NARMAX models with output multiplicities. AIChE J 44:131–140
60. Eskinat E, Johnson SH, Luyben WL (1991) Use of Hammerstein models in identification of nonlinear systems. AIChE J 37:255–268
61. Wigren T (1993) Recursive prediction error identification using the nonlinear wiener model. Automatica 29(4):1011–1025
62. Ljung L (1999) System identification. theory for the user., 2nd edn. Prentice Hall, New Jersey
63. Deistler M (2002) System identification and time series analysis: Past, present, and future. In: Pasik-Duncan B (ed) Stochastic theory and control, lecture notes in control and information sciences, vol 280. Springer, Berlin Heidelberg, pp 97–109
64. Aström K, Bohlin T (1965) Numerical identification of linear dynamic systems from normal operating records. In Proceedings of IFAC symposium on self-adaptive systems, Teddington, UK

65. Box G, Jenkins G, Reinsel G (1994) Time Series Analysis. Forecasting and Control, 3rd edn. Prentice Hall, Englewood Cliffs
66. Labat D, Ababou R, Mangin A (1999) Analyse en ondelettes en hydrologie karstique. 2e partie: Analyse multirésolution croissée de pluie-débit de sources karstiques. Comptes Rendus de Académie des Sciences. Geoscience. 329:881–887
67. Labat D, Ababou R, Mangin A (2000a) Rainfall-runoff relations for karstic springs. part I: convolution and spectral analyses. J Hydrol 238:123–148
68. Dreiss S (1982) Linear kernels for karst aquifers. Water Resour Res 18(4):865–876
69. Dreiss S (1983) Linear unit-response functions as indicators of recharge areas for large karst springs. J Hydrol 61:31–44
70. Dreiss S (1989) Regional scale transport in a karst acuifer. 2. linear systems and time moment analysis. Water Resour Res 25(1):126–134
71. Padilla A, Pulido-Bosch A (1990) A reservoir model to simulate karstic aquifers. In: International conference on calibration and reliability in groundwater modeling, ModelCARE90, pp. 323–333
72. Denić-Jukić V, Jukić D (2003) Composite transfer functions for karts aquifers. J Hydrol 274:80–94
73. Hudgins L, Friebe C, Mayer M (1993) Wavelet transforms and atmospheric turbulence. Phys Rev Lett 71(20):3279–3282
74. Labat D, Ababou R, Mangin A (2001) Nonlinearity and nonstationarity in nainfall-runoff relations for karstic springs. Institut de Mécanique des Fluides de Toulouse. Laboratoire Souterrain de Moulis. The International Association of Hydraulic Engineering and Research
75. Maraun D, Kurths J (2004) Nonlinear analysis of multivariate geoscientific data - advanced methods, theory and application. Department of Physics, Potsdam University, D-14415 Potsdam, Germany
76. Van Milligen B, Sànchez E, Estrada T, Hidalgo C (1995) Wavelet bicoherence: a new turbulence analysis tool. Physics Plasmas 8:3017
77. Onorato M, Salvetti M, Buresti G, Petagna P (1997) Application of a wavelet cross-correlation analysis to dns velocity signals. Eur J Mech 16(4):575–597
78. Labat D, Ababou R, Mangin A (2000b) Rainfall-runoff relations for karstic springs. part ii: continuous wavelet and discrete orthogonal multiresolution analyses. J Hydrol 238:149–178
79. Buresti G, Lombardi G (1999) Application of continuous wavelet transforms to the analysis of experimental turbulent velocity signals. In: Banerjee S, Eaton JK (eds) Turbulence and Shear Flow Phenomena, Begell House Inc, pp. 762–772
80. Sello S, Bellazzini J (2000) Wavelet cross-correlation analysis of turbulent mixing from large-eddy simulations. Eighth European Turbulence Conference, Barcelona pp. 27–30
81. Liu PC (1994) Wavelet spectrum analysis and ocean wind waves. In: Foufoula-Georgiou E, Kumar P. (eds.) Wavelets in geophysics pp. 151–166
82. Barrett JF (1964) Hermite functional expansions and the calculation of output autocorrelation and spectrum for any time-invariant nonlinear system with noise. J Electron Control 16: 107–113
83. Beven J (2001) Rainfall-Runoff modelling. the primer. Wiley, Chichester
84. Krause P, Boyle D, Base F (2005) Comparison of different efficiency criteria for hydrological model. Adv Geosci 5:89–97
85. Nash J, Sutcliffe J (1970) River flow forecasting through conceptual models. part i. a discussion of principles. J Hydrol 10:282–290
86. Legates D, McCabe G (1999) Evaluating the use of "goodness-of-fit" measures in hydrologic and hydroclimatic model validation. Water Resour 35:233–241
87. Donoho D, Johnstone I (1994) Ideal spatial adaptation by wavelet shrinkage. Biomefrika 81:425–455
88. Donoho D (1995) De-Noising by soft-thresholding. IEEE Trans Inf Theory 41(3):613–627
89. Bendat J (1990) Nonlinear system analysis and identification. Willey, New York
90. Zhang Q (1997) Using wavelet network in nonparametric estimation. IEEE Trans Neural Network 8(2):227–236

91. Glentis G, Koukoulas P, Kalouptsidis N (1999) Efficient algorithms for volterra system identification. IEEE Trans Signal Process 47(11):3042–3057
92. Mirri D, Iuculano G, Traverso P, Pasini G, Filicori F (2003) Non-linear dynamic system modelling based on modified volterra series approaches. Measurement 33:9–21
93. Giri F, Bai E (2010) Block-oriented Nonlinear System Identification. In: Control and Information Sciences (Lecture Notes). Springer, Heidelberg
94. Wills A, Schön T, Ljung L, Ninness B (2013) Identification of hammerstein-wiener models. Automatica 49:70–81
95. Narendra KS, Gallman PG (1966) An iterative method for the identification of nonlinear systems using a Hammerstein model. IEEE Trans Autom Control 11(3):546–550
96. Bai E, Kullback DL, Leibler RA (2004) Convergence of the iterative hammerstein system identification algorithm. IEEE Trans Autom Control 49(11):1929–1940
97. Chang F, Luus R (1971) A noniterative method for identification using hammerstein model. IEEE Trans Control Syst Technol 16(5):464–468
98. Bai E-W (1998) An optimal two stage identification algorithm for hammerstein-wiener nonlinear systems. Automatica 34(3):333–338
99. Bai E (2002) A blind approach to the hammerstein-wiener model identification. Automatica 38:967–979
100. Goethals I, Pelckmans K, Suykens J, De-Moor B (2005) Subspace identification of hammerstein systems using least squares support vector machines. IEEE Trans Autom Control 50(10):1509–1519

Chapter 3
Karst and Glacial Hydrology

Now the sun, moving as it does, sets up processes of change and becoming and decay, and by its agency the finest and sweetest water is every day carried up and is dissolved into vapour and rises to the upper region, where it is condensed again by the cold and so returns to the earth. This, as we have said before, is the regular course of nature.

Aristotle (fl. c. 384 BC–c. 322 BC)

3.1 Introduction

Water elapses throughout the Earth in forms of well known cyclic processes as evaporation of the oceans, river discharge and rainfall. Nevertheless, the cycle has some aspects that are less known, such as the hydrological cycle involving the karst groundwater flow. Also, the hydrological aspect that implicates the discharge yielded by glacier melting is among the main topics of the water cycle. Karst is a geological system that stores water from rainfall and later releases it through the spring drainage. By analogy, this phenomenon can be used to introduce as well the glacier hydrology. In this natural system, storage is determined by the persistence of the water in a solid state.

The objective of this chapter
The aim is to explain conceptually both karst and glacier systems from the hydrological point of view, so as to establish analysis requirements and remark on certain restrictions including details about the system structure as well as the physical laws that sustain their behavior and cannot be fully known. The above line of inquiry will lead to the proposal of certain data-driven model techniques.

D. Chinarro, *System Engineering Applied to Fuenmayor Karst Aquifer (San Julián de Banzo, Huesca) and Collins Glacier (King George Island, Antarctica)*, Springer Theses, DOI 10.1007/978-3-319-08858-7_3

Outline of this chapter

Section 3.2 explains the origin of karst landform and the karstification process. The classification of a karst process according to morphology is presented in Sect. 3.2.2. Section 3.2.1 describes speleogenesis through the dominant reaction in karst processes.

Section 3.2.3 establishes the main concepts in karst hydrology, with classical conceptual schemes defined in Sect. 3.2.3.1. The complexity in modeling the karst system is already mentioned in Sect. 3.2.4, with the several fundamentally different types of modeling approaches for quantitatively characterizing the hydraulic behavior of karst hydrogeological systems being discussed. Section 3.2.4.1 is a review of methods and types of recharges, where the evapotranspiration should be calculated to estimate the effective rainfall from the total precipitation (3.2.4.1) Sect. 3.2.4.2 presents the aquifer discharged through a small number of large springs.

Section 3.2.5 is a study to set up a schematic conceptual model of a real system with the aim of allowing for characterization following the methodology of system identification.

Section 3.2.6 is a overview of the main techniques of system identification and analysis implemented in karst systems that have been revised in Chap. 2. Section 3.2.6.1 reveals a strong and fast relationship between rainfall events and flow rising and indicates the aquifer inertia (system memory). A review of the study of correlation and cross-spectral analysis makes up the Sect. 3.2.6.2. Applications in diverse papers using linear, nonlinear and wavelet methods are presented in Sects. 3.2.6.3, 3.2.6.4, and 3.2.6.5.

Section 3.3 provides background on glaciers with a general description of glacier dynamics. Section 3.3.1 introduces the physical basis for a distributed energy balance in melting processes. Section 3.3.2 is a discussion about a glacier classification on thermal regime of the glacier layer. Section. 3.3.3 is a study of glaciers from a hydrological perspective, as an important water resource that contributes to streamflow.

The approaches of the most important models are briefly described in Sect. 3.3.4. A discussion about what are the main input variables involved in the glacier system appears in Sect. 3.3.4.1 and the relevance of the glacier streams in the discharge is treated in Sect. 3.3.4.2. Thereafter, the fundamentals of the two main melting models, the energy balance model (3.3.4.3) and temperature-index model (3.3.4.4), are revised.

Section 3.3.5 is an account of time series analysis and models applied to glacier hydrology, assuming a characterization of linear (3.3.5.1) and nonlinear (3.3.5.2) system structures, with special mention of wavelet techniques (3.3.5.3).

Section 3.4 is the conclusion and emphasizes the main findings arrived at in this chapter.

3.2 Karst Background

Karst word is referred to the German expression that is stemmed from the Slovene word *kras* and the Italian *carso*, and means *rocky terrain*. Karst is a landform originated primarily by dissolution on limestone, with diverse shapes, blind valleys, caves and singular springs. The term was originally used to describe a landscape in the western part of Croatia and Slovenia, where this kind of terrain was firstly studied and formally defined. Currently, the name is internationally accepted to designate similar processes anywhere in the world.

Over decades, geomorphology and hydrology of karst was separately studied as different issues. However, from 1980 both perspectives are regarded together as this karst definition explains: "Karst is a terrain with distinctive hydrology and landforms arising from the combination of high rock solubility and well developed solutional channel (secondary) porosity" [1]. An emerging approach to karst hydrogeology, gives a perspective more integrative and universal by encompassing the whole range of karst processes as this declaration: "An integrated mass-transfer system in soluble rocks with a permeability structure dominated by conduits dissolved from the rock and organized to facilitate the circulation of fluid" [2].

In order to understand the groundwater aspects, the next section starts with a brief introduction to the karst processes.

3.2.1 *Karst Processes*

Cvijic [3] is considered the pioneer in understanding the genesis of the karst cavities. His work provides a relation between the existence of karst landscapes and the groundwater flow, describing typical vertical development zones (dry, transition and saturated). He stated that the dissolution of the rocks was the key process for the formation of caves and sinkholes. Later, Ford [4] elucidates the karst formation process, though it is more detailed in the publication of Ford and Williams [5].

Karstification mainly consists of a weathering due to dissolution in water, whose aggressiveness is characterized by the action of other chemical processes (hydration, ionic substitution and oxidation-reduction), and physical phenomena (mass transfer and diffusion). Weathering is defined by Summerfield [6] as "the adjustment of the chemical, mineralogical and physical properties of rocks in response to environmental conditions prevailing at the Earth's surface". Parameters, that in general determine the weathering process, are: the type of geological material, the atmospheric pressure, temperature, and contact of water with air. The weathering process is part of the carbon cycle, in which carbon is exchanged between the atmosphere, surface water, groundwater and carbonate minerals.

Limestone, the most common sedimentary rock in karst formations, is mainly composed of the mineral calcite, the most stable polymorph of calcium carbonate ($CaCO_3$). A carbonate rock with at least by a 90 % of calcite is considered a pure carbonate rock [7]. The limestone is produced by chemical precipitation and the accumulation of organisms such as corals and gastropods. The dissolution of calcium carbonate by carbon dioxide in aqueous solution is the dominant reaction in karst processes, including speleogenesis. The reaction can be represented by:

$$CO_3Ca \rightarrow CO_3^{2-} + Ca^{2+}\Gamma \tag{3.1}$$

Along with the water, CO_2 plays an important role in the karstification process. It comes generally from the atmosphere, though also CO_2 can be produced biologically in the soil.

$$CaCO_3 + CO_2 + H_2O \rightarrow Ca^{2+} + 2HCO_3^{2-} \tag{3.2}$$

The reaction is reversible. The solution, containing the dissolved calcium bicarbonate, can lose carbon dioxide to the atmosphere and precipitate calcium carbonate. This process is responsible for the development of speleothems, and tufa or travertine at the surface [8].

Precipitation of dissolved carbonate minerals is accompanied by the release of the carbon as CO_2. The dissolution progresses toward deep underground areas with aggressive water filtration through multiple channels such as potholes, sinks, joints and chasms. In the vertical penetration, the water is away from contact with atmospheric CO_2, then the water aggressiveness lessens, losing the ability to dissolve.

3.2.2 Karst Classification

The classification of karst is very important from the standpoint of hydrogeology. Cvijic [3] provided one of the first classifications of karst according to the degree of karstification: Holokarst (complete karst) and Merokarst that defines an imperfect karst topography.

Ford and Williams [1] classify karst process according to morphology, in endokarst and exokarst, although the author admits that there may be mixed formations.

- Exokarst. All features that may be found on a surface karst landscape, ranging in size from tiny karren to large poljes, and from protruding forms to deep depressions, belong to the exokarst [8]. Exokarst can be prominent forms as *Karren* or depressions as *Doline* (a concave depression in limestone) and *Polje* that is a large depression in a karst which follows the main geological structure trends.
- Endokarst. Endokarst is the layer of the karst system beneath the surface. It is created by surface waters that slowly percolates dissolving the limestone, creating deep crevices, galleries and large caves, where water flow arises to constitute the underground runoff.

3.2.3 *Karst Hydrology*

An aquifer (from *aqua* = water and *fero* = carry) is all geological formation saturated in water that is capable of acquiring, maintaining and passing water through geological formations. Accordingly, the aquifer comprises in a broad sense, a recharging surface, a vertical flow of water through the vadose zone, a storage and transmission, and a discharge area that can be very reduced. There are three types of aquifers: detrictic, fissured and karst. The last is the typical of the limestone terrains.

From the notion of system, Yevjevich [9] firstly used the concept of the hydrological system. Mangin [10], leading the *Laboratoire souterraine of Moulis* at the French Pyrenees, translated it to the concept of "karst system". He defined karst system as *the entity where the water flow constitutes a drainage unit* and introduced an important paradigm shift on the study of karst hydrology, linking together karst genesis, morphology, hydrology and hydrochemistry. In practice, many authors use karst system as synonymous of karst aquifer, and also in this work.

Bakalowicz [11] described the various types of recharge and functions in the vadose and phreatic zones of a karst system. Karst aquifer, sometimes visitable, can exist even when there are no recognizable karst landforms on the surface, and even when there are no known and accessible caves.

The knowledge of each aquifer requires a study of recharge, internal constitution, water flowing through the rocks, the chemical and physical processes between water and rocks, and finally the process by which the aquifer is discharged, either naturally through springs and rivers, either by artificial procedures as pumped wells.

3.2.3.1 Karst Conceptual Scheme

Since the first karst studies, there was a controversy on the existence of a saturated zone. The French school, headed by E. Martel, denied the existence of a saturated zone, based on their caving experience in the Pyrenean underground rivers [12]. The Austrian school, from the study of Slovenian karst, recognized the existence of underground water zones [3]. Along the time, the ideas of both schools were going mingling till to converge, according to Ford and Williams [1].

Underground water, as such, is hardly seen until it ceases to be underground. For this reason, it is difficult to study and measure [11]. The phenomenon of the progression of the water flow in a karst system is complicated and apparently not completely known.

In the literature, most of the authors [10, 13] agreed, though differing in the terms used, to divide the karst system into several parts or subsystems to tackle the model study. Essential components of a karst aquifer and system structure were stated by White [14] and sketched in Fig. 3.1.

Currently, most conceptual models distinguish three main zones—besides superficial recharge area—in the vertical direction, taking into account the essential hydrogeological characteristics described by Mangin [10]:

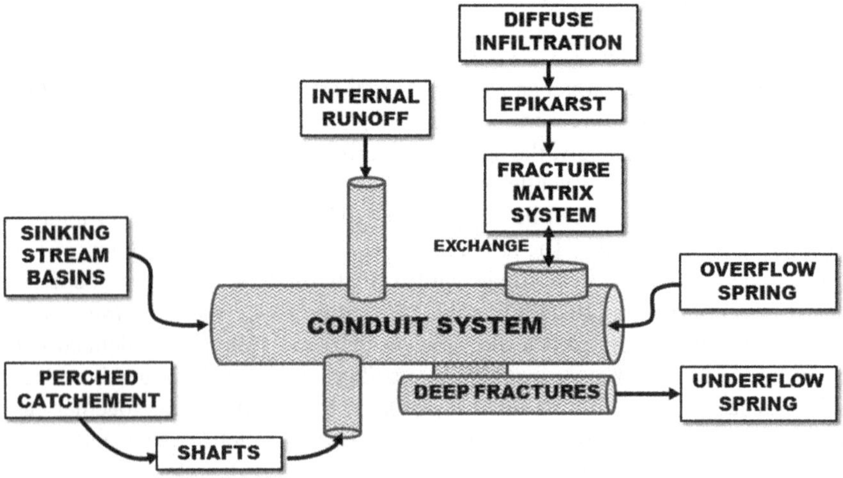

Fig. 3.1 The essential components of the karst aquifer [14]

- *Epikarst zone*, with a relatively thickness that may vary significantly (15–30 m may be a good generalization), is the portion of bedrock that extends from the base of the soil zone and is characterized by extreme fracturing and intense solution. It is separated from the phreatic zone by an inactive, relatively waterless interval of bedrock that is locally breached by vadose percolation. The epikarst develops a higher activity, due to the proximity to the surface, where water has a higher carbon dioxide concentration. Significant water storage and transport are known to occur in this zone [2].
- *Transit zone* It is below the epikarst, with which it forms the so-called **unsaturated zone** and is more defined by vertical lines that provide a rapid progress infiltration reaching the water table or piezometric.
- *Saturated zone* or *phreatic zone* comprises those parts of the karst in which all voids are filled with water. The saturated zone is formed by conduits of large dimensions, narrow cracks and pores in the rock. In many karst conduits are organized in a hierarchical network similar to a river system. This network finishes at the surgence area before commented.

In the conceptual model proposed by Mangin [10], the main conduit system transports infiltration waters towards a karst spring, but is poorly connected to large voids in the adjacent rocks, referred to as the "annex-to-drain system". The particularity of Mangin's conceptual model is that it associates the storage function of karst aquifers to the "annex-to-drain system".

The conceptual model proposed by Drogue [15] assumes that the geometric configuration of karst conduit networks follows original rock fracture patterns with a size in the order of several hundred meters, separated by high-permeability low storage conduits.

The conceptual model proposed by király [16] combines both Mangin and Drogue models. Although KirÃ¡lys' model employs a hierarchical conduit network similar to the model of Mangin and takes the effect of the epikarst into account, it also comprises the hydraulic effect of nested discontinuity groups similar to those involved in the model of Drogue. KirÃ¡ly demonstrated the scale effect to be a consequence of coexisting discontinuity groups of different scale associate it to the low-permeability matrix.

During last decades of twenty century, an important number of studies were accomplished on the karst hydrology. The result was the identification of several types of karst aquifers. White [17] proposed a conceptual scheme for classifying carbonate aquifers in terms of groundwater flow system and hydrogeologic setting, recognizing three principal flow type: diffuse, with slow laminar flow; free, with fast and turbulent flow in conduits; and confined, with generally slow flow in thin layers intercalated between impervious rocks.

3.2.4 Modeling Approaches

The complexity in modeling the karst system, already mentioned by Lastennet and Mudry [18], is determined by the high capillarity in the water flow scattered through ducts of an internal structure that is not well known [11].

Attempts to model karst aquifers using spatially distributed numerical models proved acceptable results on a bounded local scale for management purposes [19, 20]. Nevertheless, this type of numerical model requires quite large databases (permeability, porosity, transmissivity, piezometry, etc.) and computing complexity, specifically when the system is heterogeneous, as the most karst aquifers are.

There are two fundamentally different types of modeling approaches for quantitatively characterizing the hydraulic behavior of karst hydrogeological systems [21].

Global models consist of the mathematical analysis of spring discharge time series (hydrographs). According to this approach, karst systems can be considered as transducers that transform input signals (recharge) into output signals (discharge). As the acquisition of spring discharge data is relatively simple, these models have been already used since the beginning of the last century. However, global models do not take into account the spatial variations within the aquifer. Consequently, they cannot provide direct information concerning aquifer hydraulic parameter fields.

Distributive models are suitable for the quantitative characterization of the spatial variations of hydrogeological phenomena. Distributive methods consist of subdividing the model domain into homogeneous sub-units, and calculating groundwater flow by applying flow equations derived from basic physical laws.

3.2.4.1 Input: Recharge

In general the aquifer recharge is the volume of water entering the aquifer, that mostly can be estimated by the relative percentage of precipitation in the basin, which constitutes the system input signal. Recharge can be estimated by different methods. For example, Andreo et al. [22] develop a simple method to estimate the rate of carbonate aquifer recharge (expressed as a percentage of precipitation) by combining different variables (geological, geographical, morphological and soil), and to set the zonal distribution of recharge aquifers.

There are two basic ground-water recharge types in karst terranes: autogenic (concentrated or diffuse) and allogenic [23].

- Autogenic recharge is derived directly from precipitation over a closed depressions (i.e. gorges, poljes or canyons), and enters the rock at discrete points (sinkholes or large fractures). The autogenic recharge is called diffuse when there is a slow percolation through numerous small orifices and in a large area of limestone outcrop.
- Allogenic recharge to karst aquifers occurs when the surface runoff is on large areas of insoluble rock or low permeability soils, so the water flows directly from the adjacent soluble carbonate bedrock [24].

Effective rain. Rainfall is the only input of autogenic recharge in the karst unit, but the value of evapotranspiration should be discounted from the rainfall measured. This value is defined as *effective rain*. Jemcov [25] assumed the effective infiltration as the input function instead of precipitation in order to take into consideration various processes in air, vegetation and soil and their influence on recharge, and to separate them from the processes within the karst aquifer system.

In order to adequate the measured raw rainfall to the recharge area characteristics, the evapotranspiration has been taken account to get the effective rainfall. The rainfall (R) is used to soak up the overstorey, understorey and litter/moss zone (S) until a threshold (given for each type of soil) and for the evapotranspiration process (E_{TP}). The remaining water is the percolated flow inward the unsaturated zone, i.e., the Effective Rainfall R_e. The relation is given by the expression:

$$R_e = R - (S + E_{TP}) \tag{3.3}$$

Evapotranspiration (evaporation + transpiration) is a term used to describe all the processes by which liquid water becomes vapor, from the surface of water, snow field or land (evaporation) or through plant stomata (transpiration) to the atmosphere. Therefore, the calculation of the actual evapotranspiration losses are necessary in the estimation of recharge due to precipitation. Direct measurement of evapotranspiration losses is difficult, because of that the interdependence between different components of the soil-vegetation-atmosphere system hinders the proper estimation of the evapotranspiration losses [26]. There are different methods to calculate the evapotranspiration [27]. Most measurements are indirect and spatially bounded.

The choice of the method for the calculation of potential evapotranspiration in this thesis was governed by considering sufficient degree of reliability and meteorological information, required by the formula. To estimate the evapotranspiration in the karst analyzed for this thesis, the expression from Eagleman [28] has been selected:

$$E_{TP} = f.C.W.E_{\max}\sqrt{100 - H_r} \quad (3.4)$$

where, $E_{\max} = 6.1e^{(17.1T/(234.2+T))}$; f is a correction coefficient, its value is 0.0329 for the daily time step and 0.0329/24 for hourly time step; W depends on speed wind (for a calm wind is 0.8); H_r is the relative humidity; and C depends on air temperature T according to Table 3.1.

3.2.4.2 Output: Discharge

A distinguishing feature of karst aquifers is that most of the groundwater is discharged through a small number of large springs. In some karst aquifers, the entire discharge is through a single spring. One or few of these springs are the base-flow discharge and are called underflow springs, while other springs flow only during periods of high discharge and are called overflow springs (or trop plein in French) [29]. A lot of small tributaries converge inside the aquifer, to form a flow path that drains toward the surface through a spring. So, the discharge from karst springs is a composite of all water moving through the aquifer. The spring (or springs), therefore, is an optimum location for measuring some parameters that are the outputs of the karst system.

The common way to study the discharge of one spring as a function of time is the *hydrograph of a single event*, with a falling limb named *recession curve* that occurs after rainfall has ended. The integral of the curve of a hydrograph is the *flow-mass curve* that represents a hydrologic quantity of the aquifer.

The shape of the recession limb of the hydrographs of karst springs was studied from the end of the 19th century. Maillet [30] pioneered the work using a first-order LTI model approach for the drain of a single reservoir. He defined the recession coefficient alpha in the Eq. 3.5, where the discharge is proportional at any time to the water volume stored.

$$Q(t) = Q_0 e^{-\alpha(t-t_0)} \quad (3.5)$$

Table 3.1 Coefficient C in the Eagleman's expression

Coefficient	Condition
0.63	T<=0°
0.63+0.024T	0° <T<=21°
1.13	T>21

where Q_0 is the spring discharge at the beginning of the recession ($t = t_0$), and α is the recession coefficient with dimension T^{-1}. Subsequent authors introduced the technique of composite hydrograph recessions. More coefficients, sometimes empirical and without physical meaning, were developed in order to cope with the real hydrographs. Ford and Williams [31] present a summary on that issue. Many authors, e.g. [32–35] work under the hypothesis of the existence of conduits of different diameter draining at different rates. However, this approach was not satisfactory because in karst hydrogeology, the application of numerical methods requires a specially tailored methodology [24].

3.2.5 Characterization of Karst Aquifers

The first step in any model study is to set up a schematic representation (conceptual model) of a real system. The karst modelling approach chosen for this work is the *global model*, which can be characterized as a posteriori modeling or *black-box* modelling, following the methodology of system identification.

Models can be classified into these three basic models (white, gray or black). Nevertheless, differential equations, aquifer geometry, and a set of flow parameters under certain boundaries and initial conditions, are applied in the basic models to build a model adapted to a specific karst type [36].

Black box models, based in the convolution integral between the rainfall recharge and spring discharge are the simplest models for the study of karst hydrology. Black-box models are fed by the observations and with the aim of characterizing the catchment response from observed data, which determine the model structure. The mathematical relations between the input and output time series are derived without reference to physical laws. This makes them especially adequate in karst aquifers where the hydrogeological observation is limited by the great complexity and discontinuity of the medium [37]. Black box model approach is often performed using time series analysis to study the correlation and spectral analysis of the karst system [20]. The input time series is the effective rainfall over the catchment, and the output time series is total discharge through springs.

3.2.6 Survey of Techniques Applied to Karst

Below, there is a review of the main techniques of system identification and analysis, revised in Chap. 2, than have been applied in karst systems.

3.2.6.1 Correlation Analysis

The strong and fast relationship between rainfall and the rising of discharge at typical karst springs was observed by local populations since ancient times. Analytical models of groundwater processes had been developed by the end of 19th century, followed by numerical methods based partially on physical conception.

Given the heterogeneous nature of this process both in its spatial structure and in its temporal behavior, Mangin [38], in his doctoral work, concluded that it is difficult to obtain a physical model that reproduces the expected spring hydrographs in response to a specified precipitation. Mangin suggested a longer time for the autocorrelogram of time-series discharge data and a value below 0.2 indicates greater aquifer inertia (system memory). A longer time delay between zero and the peak of a cross-correlation between precipitation and discharge time-series data indicates greater transit time of a signal within an aquifer. He works at the hydrographs of three classical springs located in Pyrenees: Aliou, Baget and Fontestorbes [39] Similar methods were applied to other karst springs mostly around the Mediterranean Sea and employed to characterize different types of karst [40–46]. However, these techniques are well known in the system engineering, and they should be used very carefully following the engineering formalization of a system, with taking the output as the discharge and the input as the rainfall.

3.2.6.2 Cross-Correlation

Following the basic work of Box and Jenkins [47], several authors introduced the cross correlation and spectral analysis relating input and output in hydrology field. Examples of those papers are: [48–56].

A very complete study of correlation and cross-spectral analysis to characterize the transformations between the input function (precipitation) and the output function (discharge) was developed by Padilla and Pulido [37]. The parameters that can be deduced are the response time, the distinction between quickflow, intermediate flow and baseflow, and the mean delay. The method offers quantifiable and objective criteria for differentiation and comparisons of karst aquifers. The study of correlogram and amplitude cross functions turn out obtain the duration of the impulse response of the karst aquifer. The gain function is used to differentiate quickflow, intermediate flow and base flow. The delay in the responses can be deduced from the phase and coherence functions.

3.2.6.3 Linear Input-Output Models

Linear systems analysis has been amply applied in surface water hydrology, and scarcely in karst groundwater. Karst aquifers may be considered as time invariant given that their underground networks evolve very slowly at the human scale [57, 58], in spite of linearity is more difficult to assume. Many karst aquifers obviously have a nonlinear response stated, by example, by the functioning of trop-pleins and temporal outlets, in addition to the "normal" springs. However, linear models historically has been successfully used in the study of a large number of karst springs, and at the moment, LTI models provide powerful techniques for analysis and model building.

The linear form of the convolution integral has been widely used [58–62], who consider the karst aquifer is a LTI filter that transforms the input signal of groundwater recharge rates into the output signal of spring. Dreiss [57] presents a method of characterizing karst aquifers based on the use of single, linear kernel functions. The kernels are identified with a deconvolution technique proposed by Neuman and Marsily [63] by using estimated groundwater recharge during isolated storms and the rapid response of the system in system output. Derived kernels from data for different storms are similar in shape and time to peak in the results could be a hint to implement a prediction model. Using similar techniques, Bohm [64] developed other model for predicting the response of a spring in extreme hydrologic conditions.

A lot of works have used linear input-output models in karst aquifer analysis. Bailly-Comte [65] studied the hydrodynamic interactions between surface and groundwater in a karst aquifer. Labat [60] demonstrated that linear input-output models, based on a convolution between rainfall rate and an a priori unknown transfer function, are not very successful at reconstructing the temporal variability of karst spring flow rates, especially at small sampling intervals. In addition, Long and Derickson [66] provide a model to characterize the aquifer response to recharge by using the transfer function by deconvolution of two time series data sets-estimated recharge and the measured response. Since in the linear systems analysis, a transfer function derived from the Fourier analysis tends to be oscillatory and not physically valid, they used a logarithm function to fit and smooth the oscillatory transfer function to obtain a more representative and physically valid transfer function. The results were tested in the karst aquifer of Madison to find the time to peak response and total memory, with a logarithmic curve that showed the best fit to the system response.

Simple linear input-ouput models give, however, interesting results when applied to the description of time series structures, which may constitute a first step to karst aquifer modelling [65, 67]. In this case the input-output model is only used to describe the system without any output simulation. Denic-Jukic [68] proposed a transfer functions for karst aquifers considered as a black box model. As a result, time series analysis applied to karst flows gives information on the structure and the behaviour of a karst aquifer in an indirect way by interpreting the induced effect on the output of an input variation. Different types of measurements may be used as input and/or output: rainfall, spring discharge, runoff, water level, temperature, turbidity, electrical conductivity etc., but the main difficulty of simple input-output models comes from the choice of bivariate time series which are not influenced by a third one.

3.2.6.4 Nonlinear Models

As many karst might be not well defined and understood by lineal models, some authors were obliged to introduce different mathematical tools. In general, it is now admitted that karst systems have a non-linear and non stationary behavior [69–71], which implies that karst spring discharge does not always result from the convolution of a non time dependent transfer function and a rainfall time series.

Labat [69] studied the functional relations between rainfall and runoff based mostly on Volterra first-order (linear) or second-order (non-linear) convolution. Their work shows that the linear stochastic model, though accurate in some respects, does not represent the hydraulic behaviour of the system very well during low flow episodes and flood events. Also, Jukic [71] analyzes also the practical applicability of nonlinear kernels for the preliminary characterization of karst aquifers and the karst springs discharge modeling.

Multifractal analysis was applied by Labat [69], in daily and half-hourly rainfall rates and runoffs measured on a three French karst springs located in the Pyrenees: Aliou, Baget and Fontestorbes. They use a universal class of multifractal models based on cascade multiplicative processes, to identify multifractal subprocesses, obtaining improvements in karst springflow simulation models.

Also, there are alternative approaches based on a combination of two linear transfer functions (composite transfer functions) dedicated to so called slow and quick flows components [68], or on non-linear kernel function [60, 71].

In the last decade, new method of nonlinear system identification based on constructing neural network are emerging in publications. E.g., Lambrakis et al. [72] examined the nonlinearity and the presence of noise in the Karst spring. They applied a surrogate time series test, to reveal a high degree of nonlinearity and a deterministic nature of the spring. Nonlinear methods were employed and found to exhibit a very satisfactory predictive ability, with neural networks achieving a slightly better performance. Kurtulus et al. [73] test the ability of the artificial neural networks (ANN) to simulate input-output responses of karst aquifers, in the study of the La Rochefoucauld aquifer located in the Charente Region, western France.

3.2.6.5 Wavelets in Karst System

Spectral analysis based on Fourier transform, by itself, does not produce a satisfactory reconstitution of the observed runoff sequences and does not cover all non linear aspects in the analysis of karst systems. On the other hand, the non-linear models created on Volterra series are more accurate, but more difficult to implement [74].

The use of wavelets has some advantages over the traditional Fourier method and has been widely used in the past decade in different fields such as electrical engineering or seismic geology. In karst hydrology, the technique was introduced by the French school and expanded in other publications as ones listed below.

Labat [75] applied a method of cross-analysis of rainfall-runoff time series based on wavelet transforms (continuous Morlet wavelet transform and orthogonal multiresolution analysis), to rainfall rates and spring outflow rates measured on the karst basin of Licq-Atherey Karst System (France). The method compared to classical cross-correlation and Fourier cross-spectral analyses, appears as powerful in the study of the joint temporal variability and non-stationarity of the rainfall-runoff relationship, and it yields more precise results than Fourier cross-spectral analysis.

Since karst aquifers can be considered as inherently non-linear and non-stationary and rainfall and outflow signals appear to be non-gaussian, Labat et al. [76] suggest that this non-linearity (non-proportional effects) and non-stationarity (lack of time invariance) should be examined by a more robust rainfall-runoff model.

Wavelet analyses with both continuous and discrete orthogonal multiresolution transforms are proposed by Labat et al. [76] and applied to continuous Morlet wavelet and multiresolution analysis with data measured at different sampling rates (weekly, daily, and halfhourly). Both univariate and joint analyses are performed on hydrological signals (rainfall rates and runoffs) in order to study their essential characteristics and their relationships. They provide a pedagogical example in order to provide guidance in the interpretation of wavelet spectrum.

In order to quantify the quality of the rainfall/discharge relationship across time-scales, Labat et al. [77] propose the use of both orthogonal wavelet multiresolution analysis and cross-correlation analysis. By using the two techniques together, they show that it is possible to show, scale-by-scale, the influence of the input to the system (rainfall) on the response (discharge) of the aquifer and also to relate these results to the internal structure of the aquifer and to the degree of organisation of the karst drainage. They have applied these techniques to Aliou, Baget and Fontestorbes springs, with data along a large period of 30 years at daily sampling rate.

Mathevet et al. [78] analyze the hydrological functioning of the Bange-L'Eau-Morte karst system using classical and original techniques, recession curves, correlation and spectral analysis, noise analysis and wavelet analysis. Wavelet analyses were applied to study the non-stationary processes, in particular, phenomena precisely situated in time. This reveals useful information on the transmissivity and the degree of karstification of the system.

To characterize the transient flow regimes of the Nam La cavern conduit, northwest Vietnam, Tam et al. [79] used wavelet cross-spectrum analysis, and investigated the stationarity and linearity of the input-output transformation of the system. Wavelet analysis also highlighted the hydrodynamics of the cavern conduit for the high flow periods.

Majone [80] analyzes hydrological and geochemical signals at the springs of Prese Val and Centonia, located in two karst catchments in the Dolomiti del Brenta ridge, northeastern Italy. Both springs are sensitive to the hydraulic property variations of the karst systems. Precipitation and spring flow at Prese Val have in common strong intermittency. Wavelets multiresolution are used in geochemical analysis, under the form of electric conductivity of water, show differences in the degree of intermittency and non-stationarity reflecting the structure of the underlying karst aquifer.

Andreo [46] used a very complete time series ranged in more than a century, on rainfall, temperature and outflow of a karst spring obtained from gauging stations in the south of the Iberian peninsula. They obtained, by means of spectral and correlation analysis and continuous wavelet analysis, that the annual periodicity of the distribution of precipitation has been constant over more than 150 years, although lesser (6-monthly periods) and weaker periodicities can also be observed. Furthermore, long-term analysis reveals a rainfall periodicity of 5 (and 2.5) years, which has also been recorded in other areas of Europe due to the influence of the North Atlantic Oscillation (NAO). They compared two methodologies used in this work, correlation and spectral analysis (with different lags) and continuous wavelet transform methods, obtain similar results. and they conclude that the correlation and spectral analysis are powerful tools to detect average periodicities in a long term series of climatic and hydrological data, while the wavelet transforms show the distribution of periodic variabilities with time.

In the closer scope of the area of the current thesis, first application of wavelet correlation analysis to the karst aquifer in Guara Mountains was carried out by Chinarro [81] in the spring of Fuenmayor, San Julián de Banzo, Huesca, Spain.

3.3 Glacier Background

A glacier is a natural system capable to retain water in solid state when the input is the precipitation and the output is the runoff derived from the ice melting.

Glaciers cover about ten percent of the Earth's land surface at the present time, whereas during the Pleistocene, they covered about three times this amount [82]. Currently, the most extensive glaciers are found in Antarctica and Greenland. On the insular Antarctica, there are some glaciers and ice caps situated on South Shetland Islands. On King George island (KGI), one of the islands of this archipelago, ice covered in more than half the island surface [83]. The most important glacier on KGI, is the glacier Collins, which is the target of the analysis of this dissertation.

3.3.1 Glacier Processes

A glacier can be considered as an open system of flowing water, with an input commonly in the form of snowfall, a storage ice transformed from snow by compaction and recrystallization, and a discharge depending on the melting process. The surface temperature is the main regulator of the melting process and, thereby, of the water storage and release. Ice or water flows through the system under diverse physical actions, and leaves the system by evaporation or discharge of melting water. The balance between the rate of accumulation and the rate of melting determines the size or mass balance of the glacial system. If a glacier keeps its constant size, equilibrium is achieved in the input processes (precipitation) and deliveries at the output by ablation (discharge). If this mass balance is positive (more gain than loss), the glacier grows; if it is negative, the glacier suffers a shrinkage.

As the energy content in the climate system grows, the ice volume in glaciers and in ice caps on earth are shrunk. Fast changes in the glacier mass are due to the rising temperature trend as UNEP [84] stated. For this reason, glacier changes are recognized as high-confident indicator of climate change and as a valuable element in early detection strategies within the international climate monitoring programmes [85].

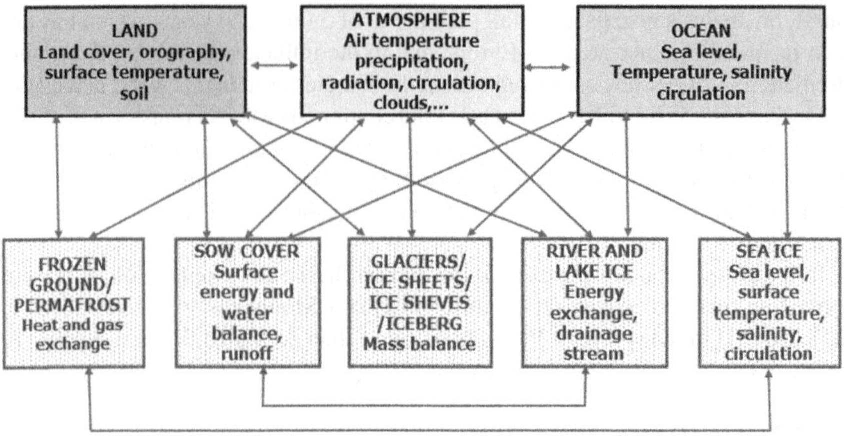

Fig. 3.2 Interface of a glacier system. The *upper boxes* indicate the state variables. *Lower boxes* indicate the processes involved in interactions oriented according the *arrows*

Nevertheless, the effects produced by the climate change could not notice immediately, because glaciers with small volume and in temperate latitude can react to climatic from a typical time scales from years to centuries [86].

Figure 3.2 illustrates how ice sheets, specially glaciers, are attached to the Earth system across its interfaces with the atmosphere, the ocean and the lithosphere [87] (Fig. 3.3).

- Glacier-ocean interface. The glacier influences the ocean through the discharge of meltwater, which influences the temperature, salinity and turbidity of the nearby ocean water. The release of ice mass, icebergs, in the nearby marine margins, influences the temperature and albedo of the surrounding ocean area. Reversely, the ocean influences the glacier. The ocean water temperature produces the ablation into submerged ice layers. Also, the ocean impacts the subglacial water pressure in the terminal zone and the wave action erodes the glacial front.
- Glacier-lithosphere interface. This interaction, among other effects, impacts the isostatic process, i.e., when the weight of the glacier depresses the Earth's mantle and causes a lithosphere deformation. After the glacier retreat, the lithosphere recovers the original shape.
- Glacier-atmosphere interface. Atmospheric variables (temperature, pressure, humidity, moisture content, energy transport) impact the mass balance and melting process. Reversely, the glacier influences the atmosphere through its high albedo, the cooling of the contact surface and the driving of atmospheric flow along the glacier course.

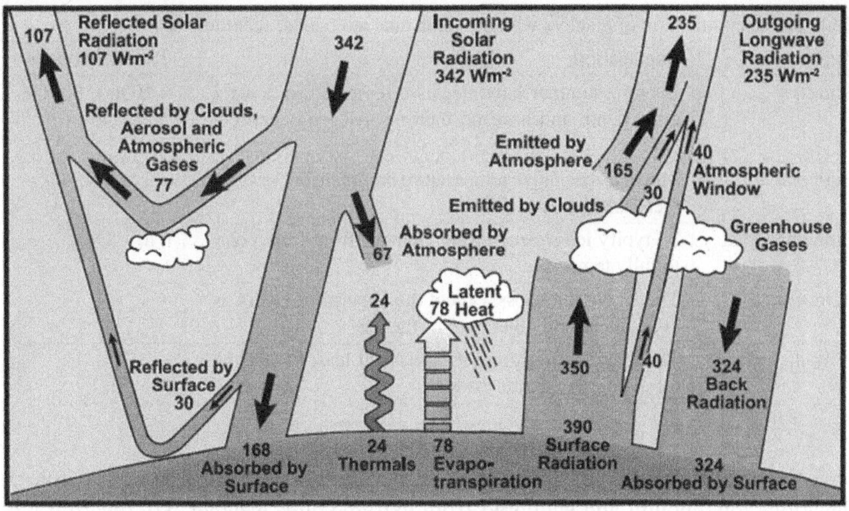

Fig. 3.3 Earth's annual and global mean energy balance, from [88]

3.3.2 Glacier Classification

A traditional classification of glaciers according to size is:

- *Mountain or alpine glaciers* are the smallest type of glacier. These glaciers can range in size from a small cirque to a larger development filling a mountain valley.
- *Piedmont glacier* is a huge mass of snow formed by coalescence of many glaciers in a plain flat region at the foot of the mountain massif. Piedmont glaciers are between several thousand to several tens of thousands of square kilometers in size.
- *Continental glacier* is the largest, with surface coverage in the order of 5 million square kilometers. Antarctica is a good example of a continental glacier.

Another classification, usually discussed among geologists, is the glacier taxonomy according to the heat state and thermal regime of the ice layer. Alhamann [89] considered the thermo-physical character of ice masses as a basis for differentiating glaciers. The glacier is tempered when its ice is at the temperature of fusion. Consequently, and based on his polar expeditions on 1930, Alhamann established two groups of glaciers: *polar* and *temperate*. About the same time, Lagally [90] presented similar propose, sub-dividing glaciers into corresponding thermodynamic categories: *kalt* (cold) and *warmen* (warm). The temperature of a polar glacier (or cold), was perennially sub-freezing throughout, except for a shallow surface which might be warmed for a few centimeters each year by seasonal atmospheric variations. On the contrary, in a temperate glacier (or warm), the temperature is at the melting point except in the uppermost layers.

Since this thermodynamic connotation, glaciers of the polar type may exist at relatively low latitudes if their elevations are sufficiently great. At the same

Table 3.2 Classification of glaciers with the main characteristics, according to [91]

Type	Explanation	Thermal bounds
Polar	Even in summer have negative temperature down a great depth and have no melting. All ice is entirely cold	−10 to −70 °C
Subpolar	Thickness negative temperature dominate but in summer melting is possible	−2 to −10 °C
Subtemperated	To typify ice sheets during changes from fully polar to fully temperate	0 to −2 °C
Temperated	The whole thickness, except the uppermost layers, is at the melting temperature of the ice	
Poly-thermal	The temperature range comprises at least two of the foregoing temperature zones	

way, temperate glaciers may be found even inside the Antarctic/Arctic Circle at elevations low enough and sheltered from extreme chill weather. So, regardless of geographical location, the internal temperature average of the glacier can be used to identify it. Though, these implications are based on changing—and usually difficult to measure—thermo-physical characteristic.

Alhamann [89] faced this problem by introducing the *sub-arctic* type—later he called it *sub-polar*—, for glaciers where the seasonal warmth penetrates at a depth significantly greater than the one experienced in summer on polar glaciers and at lesser depth than the temperated type. By the same year, Lagally also recognized this intermediate class of glacier and called it *transitional*.

To avoid some confusion from alternate application of these different terms, Miller [91] suggested including both transitional categories with other two additional types: *subtemperate* and *poly-thermal*. To elucidate the characteristics of each of these five categories, Miller suggested thermal parameters as temperature limits, which are summarized in the Table 3.2.

Miller [91] illustrates with examples the five-type classification of glaciers in the following areas: Antarctic and Greenland ice sheets (polar and poly-thermal), the Nepal Himalaya, Svalbard (polar to sub-polar), Lapland (sub-polar), sub-Arctic Norway (sub-temperate), the Alps (poly thermal to temperate), the Canadian Rockies (sub-temperate), and the Juneau Icefield, Alaska (sub-temperate to temperate). Eraso [92] claimed that glacier Collins, in the insular Antarctica and target of this study, is a temperated glacier.

3.3.3 Glacier Hydrology

From a hydrological perspective, glaciers represent important water resources, contributing significantly to streamflow. Glaciers exert a considerable influence on catchment hydrology, by temporarily storing water as snow and ice on many timescales [93].

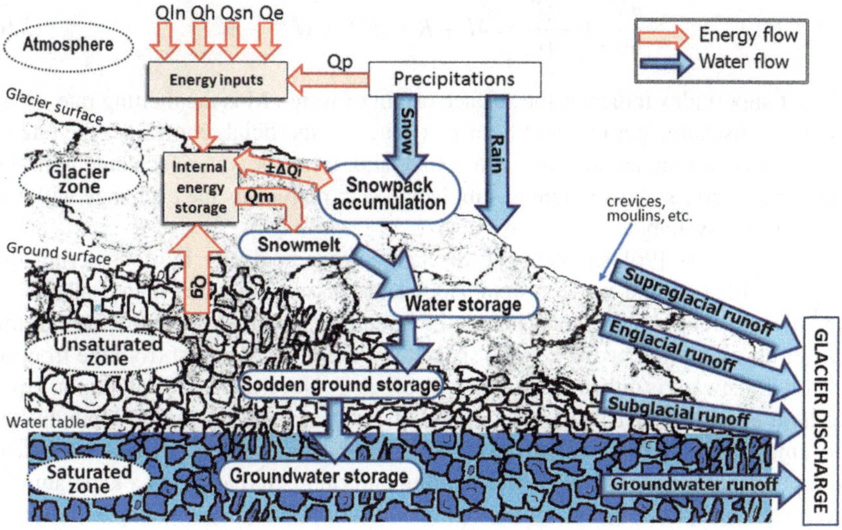

Fig. 3.4 The hydrology development and melting process in a glacier, modified from [94]

Understanding water movement through a glacier is fundamental to understand the glacier dynamics, induced floods, and the prediction of runoff throughout drainage basins. Nevertheless, glacier hydrology is a set of complex processes, since the released water suffers several changes of state and comes from several origins.

The physical structure of the snowpack is a porous matrix, and will be able to hold the liquid water between the snow grains and will increase snow density. If the snowpack voids, where the liquid water is retained, become fill up, some of the snowmelt will start to fall down and take part in some of the glacier flows. Additionally, some of the snowmelt can more deeply infiltrate into the ground to form the groundwater flows.

A simplified hydrological process development of a glacier hydrology is depicted in the longitudinal cross section of Fig. 3.4. After rain falls or snowpack melt, water can run along the glacier surface (*supraglacial* o *surface* runoff), percolates through snow by crevices and moulins to form the *endoglacier* flow, or remains dammed. An amount of this water can reach the glacier bed to set a *subglacier* flow or be retained in an inner resevoir. Finally, part of liquid can move down-gradient till the water table, to form groundwater with the common features of a hydrology underground.

A complete knowledge of a glacial hydrology comprises the total contribution of glacier surface runoff, englacial water storage and transport, subglacial drainage, and subsurface groundwater flow. The four components are represented by Flowers and Clarke [95], as a two-dimensional model with vertically integrated layers that exchange water among them. This model assumes that incompressibility and water volume are conserved according to Eq. 3.6, after simplifying some interglacial flows:

$$\frac{\partial h^r}{\partial t} + \frac{\partial Q^r}{\partial x} = M + R - \phi^{r:e} + \phi^{r:s} - \phi^{r:a} \tag{3.6}$$

where r superindex indicates the surface origin of water, M is the melting rate, Q^r is the water discharge per unit width (or flux) due to superficial runoff, R is the rate of precipitation falling as rain, $\phi^{r:e}$ is a source/sink term for water exchange with the englacial layers, $\phi^{r:s}$ is the rate to subglacial, $\phi^{r:a}$ represents the exchange with the groundwater system.

Some authors [96] simulate the discharge from a glacier by considering the analogy with a ground-water system. Hock and Janssen [97] observe two spheres of activities in studying glaciers. Though the boundary is diffuse, from the front of the glacier upward could be the general space for glaciologists, whilst from the front of glacier downward is rather an area for hydrologists. Nevertheless, the common input for both is the weather phenomenon, and under the glacier mass, the groundwater movement is similar to a karst aquifer. Figure 3.4 depicts unsaturated and saturated zones collecting water from glacier melting and rain percolation, as a karst aquifer receives water percolated from the rainfall. Both systems retain water and deliver it due to different circumstances, being the air temperature the main cause of the drainage of the glacier. This similarity in cryokarst term was presented by Eraso and Domínguez [98]. Assuming these parallel concepts, some data-driven models widely applied on karst aquifers, could be tested on glacial hydrology, with the caveat of taking in mind the glacier system singularities.

3.3.4 Modeling Approaches

The aim of models should be predicting the glacier discharge as the integral response of a glacier basin to the variable weather conditions and to show the advance or retreat of the glacier front. It is unworkable correctly calculate the impact of all the geophysical variables and meteorological factors in time and space. Therefore, a careful study of the most significant input variables of the system, could simplify the model with acceptable results. Basic model types are classified according two main categories (Fig. 3.5):

(a) **Melting models**. There is a complete hierarchy of melt models relating ablation to meteorological conditions, varying greatly in complexity and scope. Classically in glaciology two main different approaches can be distinguish in snowmelt simulation. One is well grounded in the detailed evaluation of the surface energy fluxes (energy-balance models), based on the solution of the equation of energy balance of the snow fallen and melted.

The second class of models considers the weather variables as indicators of physical processes. Concretely, the air temperature is used as unique index of melt energy (temperature-index models and degree-day method). The degree-day method has been applied in many cases for more than a century. However, these are often not practical due to large data requirements and uncertainties about spatial

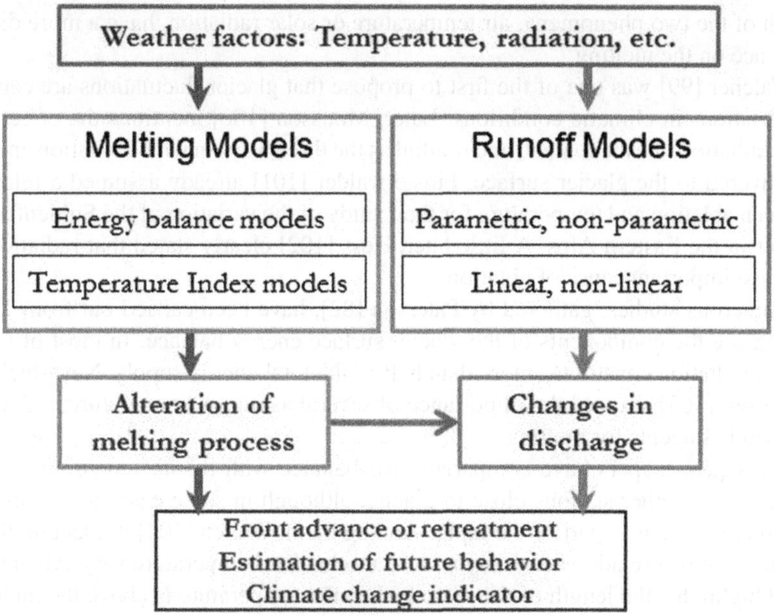

Fig. 3.5 Impact weather variables on glacier discharge, categorized by melting models and runoff models

variability. The temperature-index method, most commonly used, generally provides better performance by the low data requirements and simplicity.

(b) **Runoff models**. Among runoff models, some are rather linked to physical structures by setting equations derived from the law of mass conservation, and expressed as a balance between the internal distribution of water and external sources, e.g. [95]. Others are based on weather variables –air temperature, precipitation or radiation– as an input and the discharge of glacier-fed streams as output. The model is developed from actual daily observations of discharge and simultaneous weather conditions, taking the dynamics of glacier as a black-box system. These models provide practical solutions to hydrological problems, because of using an analytical method under a low computational cost, and requiring simple information, though long and dense time series are preferable. An example is a conceptual model identified with a transfer function or other analytic model structures, established through observations of input and output relations and constrained to a set of boundary conditions.

3.3.4.1 Input Variables

The main posed problem was to set the weight of each meteorological events which participate in the melting process. Some classic discussions have emerged to resolve which of the two phenomena, air temperature or solar radiation, have a more direct influence on the melting.

Walcher [99] was one of the first to propose that glacier fluctuations are caused by variations in climatic conditions. Later, Mousson [100] mentions the effects of solar radiation, the air contact, and rainfall as the three ways in which ablation energy is delivered to the glacier surface. Finsterwalder [101] already assumed a relation between ablation and temperature for their study of the variation of the Suldenferner glacier in the Eastern Alps. A little later, Hess [102] clearly stated that radiation is the most important cause of ablation.

Numerous studies, gathered by Paterson [82], have been carried out from 1930 to measure the components of the glacier surface energy balance. In most of them the net radiation constitutes more than half of the total energy supply. Nevertheless, Angstrom [103] stressed the importance of several agents as temperature, radiation and wind as agents for melting.

Subsequent papers have compared mass-balance with the air temperature data taken from weather stations close to glacier, although in some cases also consider the precipitation as a part of the input. Among them, Wallen [104] thinks that there is a moderate correlation between mass balance and air temperature vary. Alhamann [105] highlights the length of the period when the temperature is above the melting point, in addition other factors as the annual precipitation, the amount of incoming and outgoing radiation influenced by the degree of cloudiness, wind velocity and humidity.

Hoinkes [106] wrote: *In recent years many authors, on the basis of careful studies, have come to the conclusion that summer temperature is to be regarded as the most important factor influencing the behaviour of glaciers*. Also [106] explicitly rejects that the heat exchange from air to ice during a hot summer is sufficient to account for the greater ablation.

Many recent studies have revealed a high correlation between melt and air temperature. Braithwaite [107] confirmed that net radiation is the main energy source for melt, but the air temperature is usually better correlated with melt production and run-off than net radiation. The air temperature is used as a proxy for melt energy input to the glacier, because it has a strong relation with solar radiation [93, 108]. [107] found a correlation coefficient of 0.96 between annual ice ablation and positive air temperature sums in some Alpine glaciers.

Then the air temperature may act as the sole index of melt energy in spite of the predominance of net radiation. It is attributed to the high correlation of temperature with several energy balance components [109, 110].

3.3.4.2 Glacier Discharge

Glacier melt provides important contributions to surrounding areas through the streamflows (or glacial streams) that surge at the front of the glacier. Reversely, streamflows are a reflection of input variations of the glacier and its transformation in storage and transmission along the glacier course.

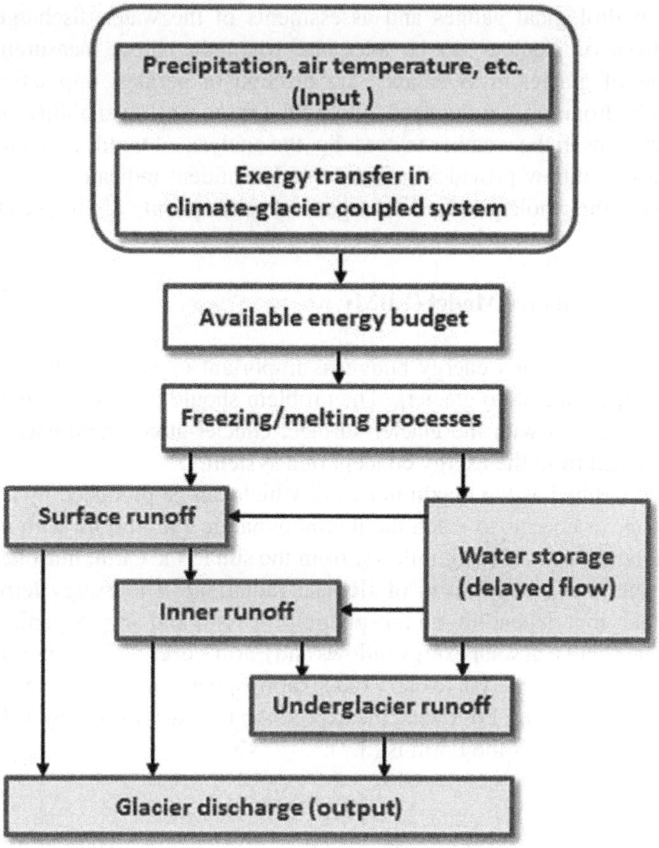

Fig. 3.6 Diagram illustrating that the exergy component due to transfer accompanying heat is the responsible of governing the melting process and consequently the discharge

Glacier melting oscillations is reflected on water discharge. The runoff has a marked diurnal peak due to the diurnal radiation cycle, with a typical delay of several hours between peak melting and peak at discharge. The annual cycle is divided between a cold and inert period, with snow accumulation, and a warmer, active period. Between the cold and warm periods, and vice-versa, there are transition seasons with complex features. The beginning is dominated by the melting of the accumulated snow, with poor water flow and refreezing processes. With the advance of the season, the melting area increases and the ice starts to melt when it is uncovered. The process continues with the formation and/or reactivation of a number of preferential vertical and horizontal flow paths on, in and under the ice, through crevasses and horizontal conduits. The melting water, routed from supraglacial, reaches the englacial and sub-glacial flow, and finally becomes the proglacial discharge (Fig. 3.6).

All the hydrological gauges and assessments of the water discharge, located after the front of glacier, can be adequate when the direct measurements and observations of glacier mass balance are difficult or perhaps impracticable. The streamflow hydrograph is influenced both by the snow mass availability and by seasonally variations in the weather factors. So, the analysis of trends in magnitude and timing in a streamflow provides sufficient and confident indicators for trends and fluctuations of the whole glacier discharge, and consequently of climate changes.

3.3.4.3 Energy Balance Model (EBM)

Knowledge of a glacier's energy budget is important to estimate the contribution of climatological factors to glaciers. The problem should be posed on how the free atmosphere interacts with the glacier surface. Glacier-atmosphere energy balance can be explained from the exergy concept of a system.

Exergy is defined as the maximum work which can be produced by a system, a flow of matter or energy, to reach the thermodynamic equilibrium with a specified reference environment. Exergy inflows, from the sun to the earth, must be balanced by exergy outflows (in the form of IR heat radiation). The exergy terrestrial inventory is varying, depending on biospheric processes and antropogenic activities. The difference between solar exergy inflows and Earth's exergy outflows is the energy available to develop work, i.e. to carry out all atmospheric, hydrospheric, geospheric and biospheric processes. Following the expression for exergy given by Valero [111], the change of exergy for the Earth is (3.7):

$$\Delta B = \int_1^2 \left(1 - \frac{T_0}{T_1}\right) \partial Q - \left[W - p_0(V_2 - V_1)\right] - T_0 \sigma; \; [injoules] \tag{3.7}$$

where T_0 and p_0 denotes the temperature and pressure at ambient conditions and T_j is the surface temperature where the heat transfer takes place. The integral term represents the exergy transfer accompanying heat, the term $\left[W - p_0(V_2 - V_1)\right]$ is the exergy transfer accompanying work, and the term $T_0\sigma$ accounts for the destruction of exergy or irreversibility.

The hydrosphere extracts exergy from the solar inflow (in the form of heat or work) in order to accumulate it in somehow, or consume it in physical processes, as the ones come about in glaciers. So, snow and ice play an important role, interacting with the atmosphere over a range of spatial and temporal scales in a complex feedback mechanisms.

Once calculated the whole exergy of a coupled system formed by climate and glacier subsystems, the exergy component due to transfer accompanying heat is the responsible of governing the melting process and consequently the discharge (Fig. 3.6).

An imbalance in the thermal exergy is an available energy budget that will affect the temperature of the glacier surface. If the glacier surface is below but near the melting point (0 °C), an additional contribution of heat transferred towards the glacier surface, will yield the melt of ice or snow. The sources energy that cause snow melt are the sum of shortwave Q_{sn} net radiation, long-wave net radiation ln, convection from the air Q_h, vapor condensation Q_e, conduction from the ground Q_g, and energy contained in rainfall Q_p.

The available energy budget Q_m is expressed by the equation (3.8), given by Price et al. [94]:

$$Q_m = Q_{sn} + Q_{ln} + Q_h + Q_e + Q_g + Q_p - \Delta Q_i \tag{3.8}$$

These contributions are usually measured as energy per time per unit area of snow. ΔQ_i is the rate of change in the internal energy stored in the snow per unit area of snowpack. In the warm period, the net flux of heat (ΔQ_i) goes inwardly the snow, while during periods of cooling, the net flux (ΔQ_i) emerges outwardly from the snowpack. If the available energy budget is sufficient, the melting process can produce liquid water. The Fig. 3.4 is a graphical synopsis of the melting process, where the energy flux is representing by red lines.

Many authors have studied the energy and mass balance of glaciers. Pioneering work concerning the details of energy exchange between the atmosphere and a glacier surface was performed in the Nordic countries. Alhamann [112] derived the first empirical formula for the computation of ablation from known values of incident radiation, air temperature and wind velocity. Sverdrup [113] computed a complete energy balance. In subsequent decades, gradient flux techniques to ice and snow were studied [114–116]. Others consider the mass balance depends strongly on altitude, [107, 115, 117]. Others have modeled mass balance gradients and sensitivities [114, 118–120]. Only a few authors have studied the surface of a glacier in a three-dimensional way [121, 122].

3.3.4.4 Temperature-Index Model (TIM)

This method computes snowmelt only by using air temperature as an index to melting process. A thermal *index* is a statistical estimator used instead of physical variables. This statistical method removes many of the variables from the equations, because is often difficult to assess. Moreover, air temperature is assumed as a predominant variable in the energy budget equations, because it is connected with many of the energy exchanges involved in snowmelt.

The basic equation for the temperature index solution is defined in (3.9), although Gray and Prowse [123] provide some derived expressions.

$$M_s = C_m(T_a - T_b) \tag{3.9}$$

where M_s = snowmelt, in. per period, C_m = melt-rate coefficient that is often variable, in./(degree/period), T_a = air temperature in °T, and T_b = base temperature in °F.

Degree-day model is a form of temperature index models that are based on an assumed relationship between ablation and air temperature usually expressed in the form of positive temperature sums. Hock [124] relates the amount of ice melt, M (mm) in The Eq. 3.10, during a period of n time intervals, dt (day), to the sum of positive air temperatures of each time interval.

$$\sum_{i=1}^{n} M = DDF \sum_{i=1}^{n} T^{+} \Delta t \tag{3.10}$$

The factor of proportionality is the degree-day factor (DDF), and is expressed in mm $d^{-1}\,°C^{-1}$. Commonly, a daily time interval is used for temperature integration, although any other time interval, such as hourly or monthly can also be used for determining degree-day factors.

The Temperature-index model of Hock and Janssen [97] is an extension of the temperature-index method, where the degree-day factor is parameterized as a linear function of potential shortwave radiation. This model has been widely used and exhibits significant improvements in predictive capability over the classical temperature-index approach. Temperature-index model of Pellicciotti et al. [125] is another alternative to the classical temperature-index. The PTIM uses an arithmetic combination of terms representing the contributions of incoming shortwave radiation and air temperature to melt. The model was applied at the point scale for glaciers in the Swiss and Italian Alps and in the semi-arid central Andes of Chile. Temperature-index models have been the most common approach for melt modelling. Hock [124] gives four reasons:

- Wide availability of air temperature data
- Relatively easy interpolation and forecasting possibilities of air temperature
- Generally good model performance despite their simplicity
- Computational simplicity

These models relay on the temperature-index approach to estimate snowmelt, but do not compute the energy balance. A review of melt factors (the controlling parameters in temperature-index models) from difference studies, however, reveals large variations between regions with no obvious climatically based pattern [124]. This result hints at the issue of parameter transferability and suggests that caution should be used when applying these models globally.

3.3.5 Data-Driven Model (DDM) for Glacier Hydrology

DDM use an ample range of times series analysis techniques, which, in this thesis, are posed from a system engineering perspective to optimize the model performance.

In some cases, it is difficult or time-consuming to get models based on the ice melting process principles. Data-driven models, or black-box mode approaches, on the other hand, are based on limited knowledge of the physical processes and depend on the information collected to characterize the functional relationship between the weather factor (mainly air temperature input) and discharge (output).

In the next sections, there are reviews and references to works already done about glacier DDM models based on discharge and temperature relationship, viewed from linear and non linear perspectives.

3.3.5.1 Linear Analysis in Glacier Systems

Few studies can be found about the glacier discharge where linear empirical models are applied. Some pioneering works are cited here given its pioneer character, or because, in other cases, they are still used due to its fast and simple implementation or as a first approach or reference for other models.

Lang [126] dealt with variations in the relation between glacier discharge and meteorological elements by regression analysis, taking in mind the specific seasonal conditions in a glacier drainage basin.

Simple linear regression discharge-temperature has been used by Braithwaite and Olesen [107] for the glaciers of Johan Dahl Land (Greenland). They calculated a correlation coefficient between (among others) the run-off and ablation (R_{qa}), which is given by Eq. 3.11:

$$R_{qa} = \frac{S_p}{S_q} R_{pa} + \alpha \left(\frac{S_a}{S_q} \right) \tag{3.11}$$

where α is the net ablation or the ratio of the total basin area and the part of basin covered by the glacier; S_q, S_p and S_a are the standard deviations of annual run-off, precipitation and net ablation respectively; and R_{qp} is the correlation coefficient between the run-off and precipitation. The R_{qa} obtained was 0.56.

Regression analysis relating daily air temperature and other variables to glacier discharge, were accomplished by Wolfe and English [127] for a glacier watershed at the Canadian Arctic. Results show that daily air temperature on the glacier is the element taken separately which shows the best correlation efficiency $R^2 = 0.57$ on the mean discharge on rain-free days, and the best prediction of glacier runoff was found using a multiple regression equation with an efficiency value of $R^2 = 0.84$, taking as input the daily air temperature, net radiation, absorptivity and wind speed.

The study of recession limbs, with the use of analogies with the emptying of one or several reservoirs, has also been used. Some papers present conceptual linear models which combine from one to three reservoirs, as what stated by Jansson et al. [93], or the work to investigate melt seasons changes in cirque glacier hydrology given by Hannah et al. [128].

Hodgkins [129] performed a correlation analysis of the time series of weather variables, in order to examine their impacts on proglacial discharge and its variation in the diurnal and seasonal cycles. The regression models developed substantially showed the best correlations of the discharge with air temperature and solar radiation.

The work from Singh et al. [130] investigates the relationships between hydrological and meteorological data collected at the front of Gangotri Glacier, Indian Himalayas. They compare of correlations between discharge and temperature, and discharge and rain of time series from five seasons, lagged by 0–3 days, to show that temperature has a better correlation with discharge series for all the years. The regression equation model used for the simulation, the efficiency obtained was $E = 0.96$.

The impact of climate variability on a set of glaciers in the northwestern Italian Alps was analyzed by Bonanno et al. [131]. The method adopted involves the use of a lagged-linear model, in which glacier snout variations depend on temperature and precipitation fluctuations. Considering the cross-correlation between these variables, they obtained a set of potential predictors related to specific seasons of the year, such as backward stepwise regression and cross-validation.

3.3.5.2 Nonlinear Models

Time dependent spectral analysis was applied by [132] to investigate seasonal variations in the second order properties of the variations of Thjorsa in Iceland and their relationships with rainfall and the melting of snow and glaciers. The first model obtained is a linear approximation to reach a highly non-linear relationship. Gudmundsson, by using Fourier power spectra, observed that the main non-linear factors, in the relationship between Thjorsa stream and the temperature, produce different effects, depending on the variations in temperature above and below the freezing point and the amount of snow on the ground.

3.3.5.3 Wavelets Techniques in Glaciology

There are few references in the literature where wavelet techniques have been applied in the process of downloading glaciers, especially in the insular Antartica. In the work of Lafrenière et al. [108], cross-wavelet spectra were applied for air temperature and discharge correlation from the two streams (one is the glacial stream draining the Wapta Icefields and another is the snowmelt-fed Bow River) in summer, to characterize runoff, temperature, and precipitation regimes and their interrelationships and inter-annual variability.

To analyze the time-scale localization of the periodical signal levels in the Patagonian proglacier lakes, Pasquini et al. [133] used continuous wavelet transform (CWT) analysis. Also, to investigate periodicities in river discharge data, Pasquini and Depetris [134] used deseasonalized monthly mean series obtained by subtracting the monthly historical means from monthly means, by applying continuous wavelet transform. By wavelet spectral analysis, Pasquini and Depetris [134] sustain the evidence that the Moreno Glacier is a stable glacier, with an oscillatory terminus, when most glaciers worldwide, due to the current global warm spell, are in open retreat.

Eraso and Domínguez [92] drive wavelet analysis in the time series from a descriptive point of view, in order to establish their structure: tendency, periodical and random components. The identification of these structures and their isolation after the factorization are used to explain the processes responsible for them. Thus, with these techniques, no hypotheses are imposed upon the series to analyse, and no previous treatment is necessary (filtering, etc). In the expeditions carried out both to temperate and subpolar glaciers in both hemispheres, these authors observed the existence of endoglacier and subglacier flows and drainages also in subpolar glaciers. With hypothesis focused on investigating the role played by subpolar glacier discharge in global warming, authors applied wavelet analysis of glacier discharge to know the temporal evolution of global warming and implementing more catchment areas in both hemispheres.

3.4 Conclusion

Although the progression of the water flow in a karst system is complicated and apparently not completely known, there are many techniques derived from physical properties and mathematical tools to establish models efficient enough to study and predict the behavior of karst aquifers.

From another point of view, karst aquifer can be considered within system engineering to be a dynamic system of the black-box variety. Recharge water, the input of the system, may originate from the karst area itself (autogenic recharge) or from adjacent non-karst areas (allogenic recharge). The set of springs is the output variable of the system. Black box models establish mathematical relations between the input and output time series without any reference to physical laws. These are useful when the hydro-geological observation is limited by the great complexity and discontinuity of the medium.

Glaciers and ice caps are located in those places where the annual amount of snowfall exceeds the amount of annual snow melt. If the snowmelt process is well understood, the hydrologic cycle, which is coupled to global climate models, can be better understood. The choice of the model follows from application objectives and available input data.

The same temperatures below pressure melting point can be found over most of a sub-polar glacier, but with the existence of small areas of template ice, near the melting point. Therefore, the evolution of these glaciers can be key indicators of ongoing climate change.

The difficulty of determining the energy budget explicitly has led to a more practical model, viz, the degree day (or more generally, temperature index) methods. The temperature-index model assumes an empirical relationship between air temperatures and melt rates over the interval of interest.

The glacier conceptual models have been examined, related to the melting process and water discharge of a glacier. The runoff model, developed from actual daily observations of discharge and simultaneous weather conditions, provides practical solutions to hydrological problems of glaciers, taking the glacier dynamics as a black-box system. Most likely, a better position in glacier melt and discharge modeling will be increasingly occupied by some of the runoff models, such as data-driven models based on spectral analysis, due to the fact they use an analytical method carrying a low computational cost and require simple information, though preferably a long and dense time series.

Studies can scarcely be found about on the relationship between the glacier or karst discharge and weather variables, where black-box models from the system identification theory are rigorously applied. For this reason, this thesis presents new approaches in the study of these natural systems using data-driven techniques, such as parametric and non parametric linear methods; as well as emerging proposals focused on non-linear identification of glacier and karst systems, as Hammerstein-Wiener models and wavelet analysis.

References

1. Ford DC, Williams P (1989b) Karst geomorphology and hydrology. Chapman & Hall, London
2. Klimchouk A (2004) Towards defining, delimiting and classifying epikarst: its origin, processes and variants of geomorphic evolution. Institute of Geological Sciences, National Academy of Science of Ukraine
3. Cvijic J (1893) Das karstphaenomen. Geographische Abhandlung 21:218–329
4. Ford D, Ewers R (1978) The development of limestone cave systems in the dimensions of length and depth. Int J Speleol 10:213–244
5. Ford DC, Williams P (2007) Karst geomorphology and hydrology. Wiley, Chichester
6. Summerfield MA (1991) Global geomorphology. Pearson Education Limited, Harlow
7. Jennings J (1985) Karst geomorphology. Basil Blackwell Ltd, Oxford
8. EPA (2002). A Lexicon of cave and karst terminology with special to environmental karst hydrology. EPA, 600, R-02, 003, (2002) Technical report. EPA, Washington
9. Yevjevich V (1959) Analytical integration of the differential equation for water storage. J Res Nat Bureau Stan B Math Math Phys 63B:43–52
10. Mangin A (1975) Contribution a l'étude hydrodynamique des aquifères karstiques. PhD thesis, Institut des Sciences de la Terre de l'Université de Dijon
11. Bakalowicz M (2005) Karst groundwater: a chalenge for new resources. Hydrol J 160:13–14
12. Martel E (1910) La theorie de grundwasser et les eaux souterraines du karst. J Cave Karst Stud 21:126–130
13. Dooge J (1973) Linear theory of hydrologic systems. Technical report, 1468. U.S. Agricultural Research Service.
14. White WB (2003) Conceptual models for carbonate aquifers. In: Speleogenesis and evolution of Karst aquifers, The Virtual Scientific Journal
15. Drogue C (1974) Structure de certains aquifères karstiques d'après les résultats de travaux de forage. Comptes Rendues Académie des Sci Paris 278:2621–2624
16. Király L (1998) Modelling karst aquifers by the combined discrete channel and continuum approach. Bull d'Hydrogéologie 16:77–98
17. White WB (2007) A brief history of karst hydrogeology: contributions of the NSS. J Cave Karst Stud 69:13–26
18. Lastennet R, Mudry J (1997) Role of karstification and rainfall in the behavior of a heterogeneous karst system. Environ Geol 32–2:114–123
19. Angelini P (1997) Correlation and spectral analysis of two hydrogeological systems in central Italy. Hydrol Sci 42(3):425–438
20. Larocque M, Mangin A, Razack M, Banton O (1998) Contribution of correlation and spectral analyses to the regional study of a large karst aquifer (Charente, France). J Hydrol 205:217–231

21. Martin J White W (2008) Frontiers of karst research. Special Publication 13, Karst Waters Institute, Leesburg, Virginia Printed in the USA
22. Andreo B, Vías J, López-Geta J, Carrasco F, Durán J, Jiménez P (1996) Propuesta metodológica para la estimación de la recarga en acuíferos carbonáticos. Boletín Geológico y Minero 115(2):177–186
23. Shuster E, White W (1971) Seasonal fluctuations in the chemistry of limestone springs: a possible means for characterizing carbonate aquifers. J Hydrol 14:93–128
24. Palmer AN (2002) A distinctly european approach to karst hydrology. Hydrol Processes 16(14):2905–2906
25. Jemcov I, Petric M (2009) Measured precipitation vs. effective infiltration and their influence on the assessment of karst systems based on results of the time series analysis. J Hydrol 379(3–4):304–314
26. El-Baroudy I, Elshorbagy A, Carey SK, Giustolisi O, Savic D (2010) Comparison of three data-driven techniques in modelling the evapotranspiration process. J Hydroinformatics 12(4):365–379
27. Brutsaert W (1982) Evaporation into the atmosphere: theory, history, and applications. Kluwer Academic Publishers, Dordrecht
28. Eagleman J (1967) Pan evaporation, potential and actual evapotranspiration. J Appl Meteorol 6(3):482–488
29. Worthington SR (2003) A comprehensive strategy for understanding flow in carbonate aquifer. In: Speleogenesis and evolution of Karst aquifers, The Virtual Scientific Journal
30. Maillet O (1905) Essais d'hydraulique souterraine et fluviale. Hermann, Paris
31. Williams P, Ford D (1989a) Karst geomorphology and hydrology. Unwin-Hyman, London
32. Trilla J, Pascual I (1974) Análisis de hidrogramas de una surgencia cárstica (Fuenmayor, Huesca). Centro de Estudios, Investigación y Aplicaciones del Agua 87:20–28
33. Milanovic P (1976) Karst hydrology and water resources, vol 1. Water Resources Publications, Colorado, pp 165–191, V. Yevjevich
34. Torbarov K (1976) In: Yevjevich V (ed) Estimation of permeability and effective porosity in karst on the basis of recession curve analysis. Water Resources Publications, Colorado, pp 121–136
35. Milanovic P (1981) Karst hydrogeology. Water Resources Publications, Colorado
36. Wheater H, Jakeman A, Beven K (1993) Progress and directions in rainfall-runoff modelling. In: Jakeman AJ, Beck MB, McAleer MJ (eds) Modelling change in environmental systems. Wiley, New York
37. Padilla A, Pulido A (1995) Study of hydrographs of karstic aquifers by means of correlation and cross-spectral analysis. J Hydrol 168:73–89
38. Mangin A (1974) Notion de système karstique. Spelunca Mémoires 8:64–68
39. Mangin A (1981b) Utilisation des analyses corrélatoire et spectrale dans l'approche des systèmes hydrologiques. Comptes Rendues Acad des Sci Paris 293:401–404
40. Freixes A, Monterde M, Morin J, Ramoneda J (1996) Recursos hídricos e implicaciones ambientales del sistema de Aiguèira (Val d'Arán). In: Recursos hídricos en regiones kársticas, pp 53–71
41. Jiménez P, Andreo B, Durán J, Carrasco F, López-Geta J, Vadillo I, Vázquez M (2001) Estudio hidrodinámico del manantial de El tempul (sierra de las cabras, cádiz, sur de españa). Bol Geológico y Min 112(2):85–102
42. Jiménez P, Carrasco F, Andreo B, Durán J, López-Geta J (2002) Caracterización de acuíferos carbonáticos del sur de España a partir de su respuesta hidrodinámica, pp 105–113. F. Carrasco and B. Andreo
43. Amraoui F, Razack M, Bouchaou L (2004) Comportement d'une source karstique soumise à une sécheresse prolongée: la source bittit (maroc). Comptes Rendues Acad des Sci Geosci 336:1099–1109
44. Pérez I, Jiménez P, Andreo B, Carrasco F (2004) Estudio de la descarga de la vertiente meridional de Sierra Tejera (Málaga) mediante análisis correlatorio y espectral. Geogaceta 35:163–166

45. Valdes D, Dupont J, Massei N, Laigne B, Rodet J (2006) Investigation of karst hydrodynamics and organization using autocorrelation and t-c curves. J Hydrol 329:432–443
46. Andreo B, Jiménez P, Durán J, Carrasco F, Vadillo I, Mangin A (2006) Climatic and hydrological variations during the last 117–166 years in the south of the Iberian peninsula, from spectral and correlation analyses and continuous wavelets analyses. J Hydrol 324:24–39
47. Box G, Jenkins G (1976) Time series analysis: forecasting and control. Holden-Day, San Francisco
48. Mangin A (1981a) Apportes des analyses corrélatoire et spectrale croisés dans la connaissance des systèmes hydrologiques. Comptes Rendues Acad des Sci Paris 293:1101–1104
49. Mangin A (1982a) L'approche systemique du karst, consequences conceptuelles et methodologiques. In Reunión monográfica sobre el karst, Larra
50. Mangin A, Pulido A (1983) Aplicación de los análisis de correlación y espectral en el estudio de acuíferos kársticos. Tecniterrae, pp 53–65
51. Benavente J, Pulido-Bosch A, Mangin A (1985) Application of correlation and spectral procedures to the study of discharge in a karstic system (Eastern Spain). In: Karst water resources (proceedings of the ankara-antalya symposium), pp 67–75
52. Cruz J, Ibarra V, Morales T (1987) Aplicación de los análisis de correlación y espectral en el estudio del hidrograma del manantial de Olalde (Vizcaya). Geolis 1:50–61
53. Pulido A, Marsily G, Benavente J (1987) Análisis de la descarga del Torcal de Antequera mediante deconvolución. Hidrogeología 2:17–28
54. Mangin A Pulido-Bosch A (1991) Identification of the overexplotation of karstic aquifers using the statistical treatment of time series-2: Applications. In: Proceedings of the XXIII I.A.H. Congress "Aquifer overexplotation", pp 69–72
55. Morales T, Antiguedad I (1992) Diferenciación de tres sistemas kársticos de Bizkaia (País Vasco) a partir del análisis de sus respuestas naturales, pp 215–231. G.T.I.C.E.K
56. Rodríguez J, Antiguedad I, Llanos H, Morales T (1994) Caracterización hidrodinámica de los acuíferos kársticos del pan de guajaibón, pinar del rio, cuba, a partir de las respuestas naturales de sus surgencias. In El Karst y los Acuíferos Kársticos, pp 137–166. Universidad de Granada, A. Pulido-Bosh and J.R. Fagundo and J.E. Rodríguez Rubio
57. Dreiss S (1982) Linear kernels for karst aquifers. Water Resour Res 18(4):865–876
58. Padilla A, Pulido-Bosch A (1993) Application of a reservoir model to simulate the Torcal de Antequera karstic aquifer (Malaga), pp 51–66. A. Pulido-Bosch, University of Granada
59. Dreiss S (1983) Linear unit-response functions as indicators of recharge areas for large karst springs. J Hydrol 61:31–44
60. Labat D, Ababou R, Mangin A (2000a) Rainfall-runoff relations for karstic springs. Part I: convolution and spectral analyses. J Hydrol 238:123–148
61. Jukić D, Denic-Jukić V (2004) A frequency domain approach to groundwater recharge estimation in karst. J Hydrol 289:95–110
62. Bailly-Comte V (2006) Approche descriptive, analyse fonctionnelle et modélisation hydrologique appliquées au bassin versant expérimental du Coulazou, Causse d'Aumelas, France. PhD thesis, Université Montpellier II. Sciences et Techniques du Languedoc Dijon, France
63. Neuman S, de Marsily G (1976) Identification of linear systems response by parametric programing. Water Resour Res 12(2):253–262
64. Bohm B (1998) A model for predicting the response of a spring to extreme hydrologic conditions: A case of study of Big Sprong, Missouri. PhD thesis, Department of Geological Sicences, University of Missouri-Columbia
65. Bailly-Comte V (2008) Interactions hydrodynamiques entre les eaux de surface et les eaux souterraines en milieu karstique. PhD thesis, Université Montpellier II
66. Long A, Derickson R (1999) Linear systems analysis in a karst aquifer. J Hydrol 219:206–217
67. Mangin A (1984) Pour une meilleure connaissance des systèmes hydrologiques à partir des analyses corrélatoire et spectrale. J Hydrol 67:25–43
68. Denic-Jukić V, Jukić D (2003) Composite transfer functions for karst aquifers. J Hydrol 274(1–4):80–94

69. Labat D, Mangin A, Ababou R (2002b) Rainfall-runoff relations for karstic springs: multifractal analyses. J Hydrol 256(20):176–195
70. Massei N, Dupont J, Mahler B, Laignel B, Fournier M, Valdes D, Ogier S (2006) Investigation transport properties and turbidity dynamics of a karst aquifer using correlation, spectral, and wavelet analysis. J Hydrol 329:244–257
71. Jukić D, Denic-Jukić V (2006) Nonlinear kernel functions for karst aquifers. J Hydrol 328:360–374
72. Lambrakis N, Andreou A, Polydoropoulos P, Georgopoulos E, Bountis T (2000) Nonlinear analysis and forecasting of a brackish Karstic spring. Water Resour Res 36(4):875–884
73. Kurtulus B, Razack M (2007) Evaluation of the ability of an artificial neural network model to simulate the input-output responses of a large karstic aquifer: the La Rochefoucauld aquifer (Charente, France). Hydrol J 15:241–254
74. Mirri D, Iuculano G, Traverso P, Pasini G, Filicori F (2003) Non-linear dynamic system modelling based on modified Volterra series approaches. Measurement 33:9–21
75. Labat D, Ababou R, Mangin A (1999b) Analyse en ondelettes en hydrologie karstique. 2e partie: analyse en ondelettes croissé pluie-débit. Comptes Rendues Acad des Sci Geosci 329:881–887
76. Labat D, Ababou R, Mangin A (2000b) Rainfall-runoff relations for karstic springs. Part II: continuous wavelet and discrete orthogonal multiresolution analyses. J Hydrol 238:149–178
77. Labat D, Ababou R, Mangin A (2002a) Analyse multirésolution crosée des pluies et débits de sources karstiques. Comptes Rendues Académie des Sciences. Geoscience 334:551–556
78. Mathevet T, Lepiller M, Mangin A (2004) Application of time-series analyses to the hydrological functioning of an alpine larstic system: the case of bange-l'eau-morte. Hydrol Earth Syst Sci 8:1051–1064
79. Tam V, Smedt F, Batelann O, Dassargues A (2004) Characterization of a cavern conduit system in Vietnam by time series correlation, cross spectrum and wavelet analyses. Hydrol Sci J 49(5):879–900
80. Majone B, Bellin A, Borsato A (2004) Runoff generation in karst catchments: multifractal analysis. J Hydrol 294:176–19
81. Chinarro D, Cuchí J, Villarroel J (2010) Application of wavelet correlation analysis to the karst spring of Fuenmayor. San Julián de Banzo, Huesca, Spain. In: Andreo B, Carrasco F, Durán JJ, LaMoreaux JW (eds) Advances in research in Karst media. Springer, Heidelberg, pp 75–81
82. Paterson W (1994) The physics of glaciers, 3rd edn. Pergamon Press, Oxford
83. Clapperton C, Sugden D, Pelto M (1989) The relationship of land terminating and fjord glaciers to Holocene climatic change, South Georgia, Antarctica. In: Oerlemans J (ed) Glacier fluctuations and climatic change. Kluwer Academic Publishers, Dordrecht
84. UNEP/WGMS (2008) Global glacier changes: fact and figures. United Nation Environment Programme and World Glacier Monitoring Service
85. WMO (2010) World meteorological organization. Technical report, United Nations Environment Programme. International Council for Science, Implementation plan for the Global observing system for climate
86. (1972) Glossary and multilingual equivalents of karst terms. United Nations Educational. Scientific and Cultural Organization, Paris
87. Boultone GS (2007) Glacier science and environmental change. Blackwell Publishing, Glaciers and their coupling with hydraulic and sedimentary processes
88. Kiehl J, Trenberth K (1997) Earth's annual global mean energy budget. Bulletin Amer Meteorol Soc 78:197–206
89. Alhamann H (1933) Scientific results of the Swedish-Norwegian Arctic expedition in the summer of 1931, Part VIII. Geografiska Annaler, pp 161–216
90. Lagally M (1932) Zur thermodynamik der gletscher. Zeitschrift für Gletscherkunde 20:199–214
91. Miller MM (1976) Thermo-physical characteristics of glaciers-toward a rational classification. J Glaciol 16(74):297–300

92. Eraso A, Domínguez M (2006) Subpolar glaciers network as natural sensors of global warming evolution. Ukranian Antarct J 4–5:272–277
93. Jansson P, Hock R, Schneider T (2003) The concept of glacier storage: a review. J Hydrol 282:116–129
94. Price A, Hendrie L, Dunne T (1979) Evaluation of a snowmelt model in a boreal forest. Controls on the production of snowmelt runoff. In: Proceedings of snow cover runoff 26–28:257–268
95. Flowers GE, Clarke GKC (2002) A multicomponent coupled model of glacier hydrology 1. Theory and synthetic examples. J Geophys Res B11, 2287, 107 (B11):2287–2305
96. Derikx L (1969) Glacier discharge simulation by ground-water analogue. In: Symposium on the hydrology of glaciers, Cambridge, pp 29–40 (publication No. 95 de l' association internationale d'Hydrologie scientifique)
97. Hock R, Janssen P (2006) Modelling glacier hydrology. Encyclopedia of Hydrological Sciences, pp 2648–2655
98. Eraso A, Domínguez C (2005) GLACKMA (Glaciers, cryokarst and environment. http://www.glackma.es/. Last online Accessed 19 July 2012
99. Walcher J (1773) Nachrichten von den eisbergen in tyrol. Kurzbocken. Frankfurt and Leipzig, p 99
100. Mousson A (1854) Die gletscher der jetztzeit. Schulthess. Bayerische staatsBibliothek münchener digitalisierungs zentrum,Technical report, Zurich, Verlag von Fr
101. Finsterwalder S, Schunk H (1887) Der Suldenferner. Zeitschrift des Deutsehen und Osterreichisehen Alpenvereins 18:72–89
102. Hess H (1904) Die gletscher. Braunschweig, F. Vieweg und sohn
103. Angström A (1933) On the dependence of ablation on air temperature, radiation and wind. Geografiska Annaler 15:264–271
104. Wallen C (1949) The shrinkage of the Karsa Glacier and its probable meteorological causes. Geografiska Annaler 31(1–4):275–291
105. Alhamann H (1953) Glacier variations and climatic fluctuations. American Geographical Society, New York (Bowman Memorial Lectures, Series 3)
106. Hoinkes H (1955) Measurements of ablation and heat balance on Alpine glaciers. J Glaciol 2(17):497–501
107. Braithwaite R, Olesen O (1988) Effects of glaciers on annual run-off. Johan Dahl Land. South Greenland. J Glaciol 34(117):200–206
108. Lafrenière MM, Sharp AK (2003) Wavelet analysis of inter-annual variability in the runoff regimes of glacial and nival stream catchments, Bow Lake, Alberta. Hydological processes 17:1093–1118
109. Ambach W (1988) Heat balance characteristics and ice ablation, western EGIG-profile, Greenland, Seventh Northern Research. Basins symposium/workshop: applied hydrology in the development of Northern Basins, May 25-June 1. Copenhagen Danish Society for Arctic Technology, Ililissat, pp 59–70
110. Kuhn M (1993) Methods of assessing the effects of climatic changes on snow and glacier hydrology. In: Snow and glacier hydrology. Kathmandu symposium. 1992 IAHS, vol 218, pp 135–144
111. Valero AC, Valero AD (2010) Exergy analysis of resources and processes. Servicio de Publicaciones de la Universidad de Zaragoza, Prensas universitaria de Zaragoza
112. Alhamann H (1948) Glaciological research on the North Atlantic coasts. The Royal Geographical Society, London
113. Sverdrup H (1935) Scientific results of the Norwegian-Swedish spitsbergen expedition in 1934. Part IV. The ablation on Ischsen's Plateau and on the 14th of July Glacier. Geografiska Annaler, pp 145–218
114. Ambach W (1963) Untersuchungen zum Energieumsatz in der Ablationszone des grönlandischen. Inlandeises (Camp IV-EGIG, 69°40′05″ N, 49°37′58″ W). Meddelelser om Grønland, 174(4):311

115. Munro D (1990) Comparison of melt energy computations and ablatometer measurements on melting ice and snow. Arctic Alpine Resour 22(2):153–162
116. Hay J, Fitzharris BB (1988) A comparison of the energy-balance and bulk aerodynamic approaches for estimating glacier melt. J Glaciol 1988:34
117. Greuell W, Knap W, Smeets P (1997) Elevational changes in meteorological variables along a midlatitude glacier during summer. J Geophys Res Atmos 102(D22):25941–25954
118. Oerlemans J (1992) Climate sensitivity of glaciers in southern Norway: application of an energy balance model to Nigardsbreen, Hellstugubreeen and Alfotbreen. J Glaciol 38(129):223–232
119. Van de Wal R, Oerlemans J (1994) An energy balance model for the Greenland ice sheet. Globaland Planet Change 9:115–131
120. Jóhannesson T (1997) The response of two Icelandic glaciers to climate warming computed with a degree-day glacier mass balance model coupled to a dynamic glacier model. J Glaciol 43(144):321–327
121. Arnold NSIC, Sharp MJ, Richards KS, Lawson WJ (1996) A distributed surface energy-balance model for a small valley glacier. I. Development and testing for Haut Glacier d'Arolla, Valais, Switzerland. J Glaciol 42(140):77–89
122. Oerlemans J, Björnsson H, Kuhn M, Obleitner F, Palsson F, Smeets CJPP, Vugts HF, de Wolde J (1999) Glacio-meteorological investigations on Vatnajökull, Iceland, summer 1996, an overview. Boundary-Layer Meteorol 92(1):3–26
123. Gray D, Prowse T (1993) Snow and floating ice. In: Maidment DR (ed) Handbook of hydrology. McGraw-Hill Inc, New York, pp 7.1–7.58
124. Hock R (2003) Temperature index melt modelling in mountain areas. J Hydrol 282:104–115
125. Pellicciotti F, Brock B, aand P. Burlando, U. S., Funk, M., and Corripio, J. (2005) An enhanced temperature-index glacier melt model including the shortwave radiation balance: development and testing for Haut Glacier d'Arolla, Switzerland. J Glaciol 51:573–587
126. Lang H (1973) Variations in the relation between glacier discharge and meteorological elements. In: Symposium on the Hydrol. Glaciers, vol 95, pp 85–94. IASH Publications
127. Wolfe P, English M (1995) Hydrometeorological relationships in a glacierized catchment in the Canadian high arctic. Hydrol Processes 9–8:911–921
128. Hannah D, Gurnell A, McGregor G (2001) A conceptual, linear reservoir runoff model to investigate melt seasons changes in cirque glacier hydrology. J Hydrol 246:123–141
129. Hodgkins R (2001) Seasonal evolution of meltwater generation, storage and discharge at a non-temperate glacier in Svalbard. Hydrol Processes 15:441–460
130. Singh P, Kumar A, Kumar N, kishore N (2010) Hydro-meteorological correlations and relationships for estimating streamflow for Gangotri Glacier basin in Western Himalayas. Int J Water Resour Environ Eng 2(3):60–69
131. Bonanno R, Ronchi C, Cagnazzi B, Provenzale A (2012) Response of Alpine glaciers in north-western Italian Alps for different climate change scenarios. In: The climate of the Mediterranean region: understanding its evolution and effects on environment and societies. MedCLIVAR 2012 Conference, Madrid.
132. Gudmundsson G (1970) Short term variations of a glacier-fed river. Tellus 22:341–353
133. Pasquini AI, Lecomte KL, Depetris PJ (2008) Climate change and recent water level variability in Patagonian proglacial lakes, Argentina. Glob Planet Change 63:290–298
134. Pasquini AI, Depetris PJ (2011) Southern patagonia's perito moreno glacier, Lake Argentino, and Santa Cruz River hydrological system: an overview. J Glaciol 405:48–56

Chapter 4
Analysis and Identification of Fuenmayor Aquifer

... a country like ours, where, because of its elevated heights average, its rivers have to pour waters tumultuously, in a country so unfortunate as ours where the cries of pain by floods drown the anguish caused by drought...

Lucas Mallada (1841–1921)

4.1 Introduction

FUENMAYOR spring has been monitored continuously for identification purposes to study the behavior of a karst groundwater system. Under a linear time invariant hypothesis, the application of the simple correlation of spectral analysis and parametric identification of the transfer function generated some interesting results in the monitored spring.

These tools have historically been successful in studying a large number of karst springs and continue to be practical approximations in initial attempts to obtain a draft model. Because of the nonlinear and nonstationary nature of karst, more effective systemic techniques are required to cover certain aspects of analysis that the linear system cannot reveal adequately.

This chapter presents interesting results using Fuenmayor spring data, collected over several years, to show the ability of wavelet techniques in the analysis and identification of a karst spring system. The essential content of this chapter has been published by Chinarro et al. [4] and Cuchí et al. [6].

D. Chinarro, *System Engineering Applied to Fuenmayor Karst Aquifer (San Julián de Banzo, Huesca) and Collins Glacier (King George Island, Antarctica)*, Springer Theses, DOI 10.1007/978-3-319-08858-7_4

The objective of this chapter

This chapter aims at system identification and analyzes the time series of effective rainfall and discharge in Fuenmayor spring, trying diverse structures in order to define linear and non linear models that characterize karst hydrology.

Outline of this chapter

This chapter is organized as follows:

Section 4.2 is the geographical framework of the area under study in Natural de la Sierra y Cañones de Guara (Natural Park of Guara Mountains and Canyons) located in the central and highest part of the Pyrenean External Sierras, where the Fuenmayor spring is situated.

An overview of the Fuenmayor spring is in Sect. 4.3, with a description of the karst system and how the water appears at the contact between the limestone of the Guara formation and bounding of the recharge area around the hills surrounding Ciano polje (Sect. 4.3.1). A hydrological overview is given in Sect. 4.3.2, where the behavior of the aquifer is known for the evolution of its spring discharge and for a strong relationship between rainfall and discharge. Moreover, a literature review on this spring is provided. Section 4.3.3 is a list of infrastructure and facilities that have supported the monitoring and exploitation of the aquifer hydrology data.

Section 4.4 offers an analysis of signals or time series study. One paragraph highlights the importance of choosing the hourly sampling period (Sect. 4.4.1) in order to avoid the loss of information, as well as issues of preparation data (Sect. 4.4.2). In order to find causality between the precipitation phenomena and the aquifer discharge, Sect. 4.4.3 presents simple correlation and spectral analysis with results about autocorrelation and the spectral density of the discharge of the Fuenmayor spring. In addition, Sect. 4.4.4 uses wavelet power spectrum to reveal cyclic behavior patterns in precipitation and discharge signals. The analysis in this section assumes the nonlinearity of this karst spring system to discuss the results from scalogram observations, proposing a method for rainfall-discharge model calibration and estimating the representation of cross power spectrum and coherence.

The key problem in the system identification is to find a model structure with sufficient flexibility to suit the karst dynamics. System identification procedures start with nonparametric identification, showing the hydrogram and recession limb (Sect. 4.5.1.1), and go on by creating a model based on cross-correlation and impulse response.

The Parametric Identification of Fuenmayor section deals with the implementation of linear methods (Sect. 4.4.3) for the study of the behavior of this spring, to obtain the first transfer function in this aquifer, proposed under the hypothesis of a linear and time invariant response.

Using previous results from linear identification and spectral analysis, the final stage in this chapter consists of tuning a nonlinear model from Hammerstein-Wiener structure.

Finally, in the Conclusion section, some important assertions are made related to linear and non-linear identification of karst springs, and how to interpret the coherence results.

4.2 Geographical Framework

Parque Natural de la Sierra y Cañones de Guara (Natural Park of Guara Mountains and Canyons) is located in the central and highest part of the Pyrenean External Sierras, forming a large limestone barrier that constitutes the southern limit of the Pyrenees of Huesca, Spain. The site, declared Natural Park by Courts of Aragon in 1990, has a surface of 47,450 ha, in addition to 33,775 ha of Peripheral Protection Zone, involves 15 municipalities, and comprises the Sierra de Guara, Gabardiella, Arangol, Balcés and Sevil. The Tozal de Guara, with 2.077 m of altitude, is the highest point of the park.

Guara mountains represents the southern part of the mantle of Gavarnie. The most important geological materials are the fissured and karstified Eocene limestone, reaching thicknesses close to one thousand meters. There are, nevertheless, several impervious units delimiting diverse hydrogeological units [29]. The thickness of Guara Formation limestone increases from west to east and is intensely fractured, which favors its dissolution. The alpine orogenic creates a series of anticlinal and synclinal structures that have provided the establishment of major structural forms and a dense network of limestone fractures [9].

During the last millions of years, an intense process of karstifications has been performed in Guara Mountains. The most important karst features are the doline field of Cupiarlo, and the poljes of Ciano and Vallimona [31].

The hydrology is directly related to karst features. The area is drained by four major rivers: Flumen, Guatizalema, Alcanadre and Vero, deeply incesed by stunning canyons (Fig. 4.1). The underground of the Guara Mountains shows sinkholes and caves produced by groundwater flow under karst conditions. In the area, several karst aquifers have been identified by their springs, for example, Trinity of Rasal, Petrolanga, Solencios of Bastarás, Santa Cilia, Morrano and Pedruel, and Fuenmayor spring. The last one is the focus of this study.

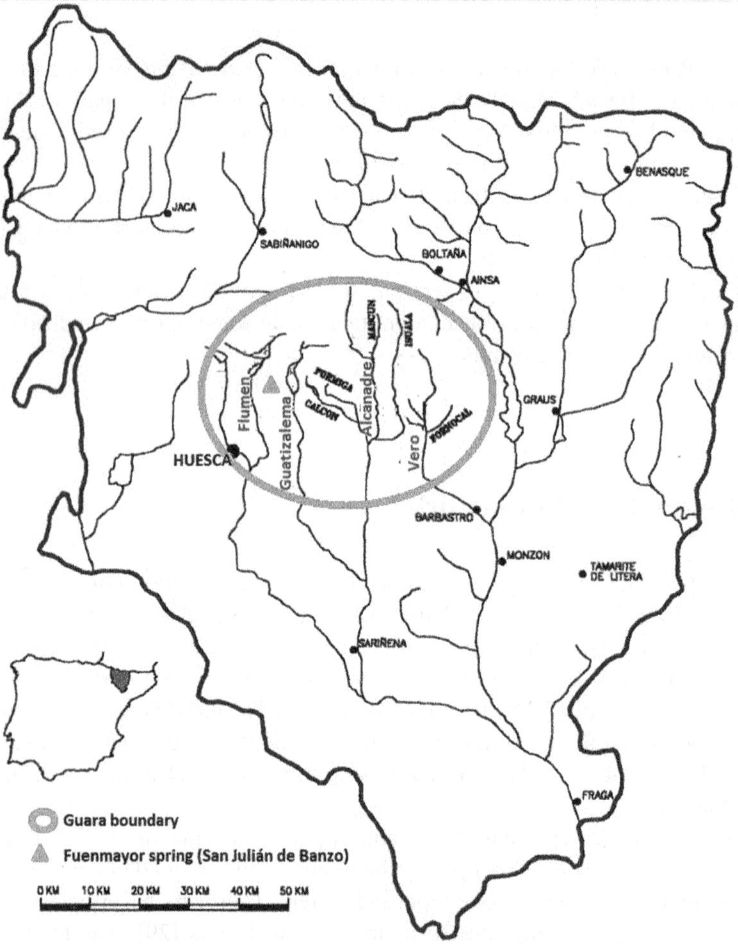

Fig. 4.1 Huesca province containing Parque Natural de la Sierra y Cañones de Guara. The green triangle is the location of Fuenmayor spring. Map from Cuchí and Setrini [11]

The area has a typical mediterranean weather with rainfall periods in Spring and Fall. The average rainfall, from 1955 to 1984, was 793 mm. The interannual precipitation shows the typical irregularity of the Mediterranean climate where droughts are common. The average temperature is near 12 degrees Celsius.

4.3 Fuenmayor Spring

The karst spring of Fuenmayor, located at 714 m (a.s.l.), emerges near San Julian de Banzo (Fig. 4.2b), which is a small village with about 50 inhabitants, located at the north of Barluenga and Chibluco, in the province of Huesca (Spain). The ravine of

Fig. 4.2 Overview of Fuenmayor spring basin. **a** Polje of Ciano, North of San Julian de Banzo. It is the northern area of Fuenmayor aquifer recharge A canyon runs dry from the small Ciano polje until the spring. **b** The headwater basin of Fuenmayor spring and an overview of San Julián de Banzo, with *Salto de Roldan canyon* on the background

San Julian comes from the spring and flows into the Flumen river; it is the axis of the Fuenmayor basin and natural separation of the two districts of the town, Suso and Yuso (Fig. 4.3).

Fuenmayor spring appears at the foothill of the External sierras of the south central Pyrenees [20]. Located about 20 km from the city of Huesca, spring has been used for drinking water supply in the city since 1880, the year in which a cholera epidemic propagated in the area near the fluvial resources which forced the seeking of other alternatives [5].

4.3.1 Geological Description

San Julian de Banzo village, located at the startup of glacis Somontano, is dominated by the escarpment of Pre-Pyrenees conglomerates formed by Salto del Roldan canyon(1,121 m) and San Martin de la Baldosera (1,431 m), and limestone of Serreta de Vallés (1,128 m) and Matapaños (1,532 m). The southern boundary is formed by sandstone and clay of *Sarinena formation* (Miocene), covered on terraces and glacis. Structurally the area is part of the inflexion in Southern Pyrenean frontal overthrusting, between San Julian de Banzo and La Almunia del Romeral. It consists of a series of folds with an east-west orientation [7]. Eocene and Upper Cretaceous limestone host a modest karst aquifer, drained by Fuenmayor, located between the back overthrusts of San Julián and Cuello Bail (Fig. 4.4).

Fuenmayor is a spring of dammed type, without trop pleins. The water appears at the contact between the limestone of the Guara formation (Middle Eocene) and the sandstone of the *Sarinena formation* (Miocene). The spring is located in the middle reach of the Molon creek. Its small canyon runs dry from the small Ciano polje (Fig. 4.2a) until the spring. Local population believed that recharge area of the

Fig. 4.3 Aerial view of the source area of Fuenmayor spring, marked by a *red circle*. In both side of the stream, in the left corner of the picture, are the districts of San Julian de Banzo (Suso to the *left*, and Yuso to the *right*). Courtesy of Google-Map

spring is located in the hills around Ciano polje. Trilla and Pascual [33] estimated a recharge area under 15 km^2. This value was agreed by Pinilla [28]. The recharge area was also calculated by Cuchí et al. [5], following a structural and hydrological criteria, to be near 10 km^2 (Fig. 4.5). Later, Oliván [25] points out an area of 970 km^2. An extra allogenic recharge may arrive from the San Martin conglomerates through the homonomous creek [10]. In the associated recharge area, the Keuper clays acted as the impervious base [28].

Near Fuenmayor there is Dos Caños spring situated about 500 meters southeast of the Fuenmayor spring (Fig. 4.5). Dos Caños drains a small aquifer with a recharge area does not exceed the square kilometer. Initially Dos Caños was considered independent of Fuenmayor in their hydrology and hydrochemistry data. Water from the Dos Caños is warmer and saltier than the water from Fuenmayor. Both aquifers are limited by the supposed impervious clays and evaporites of the Keuper facies [28].

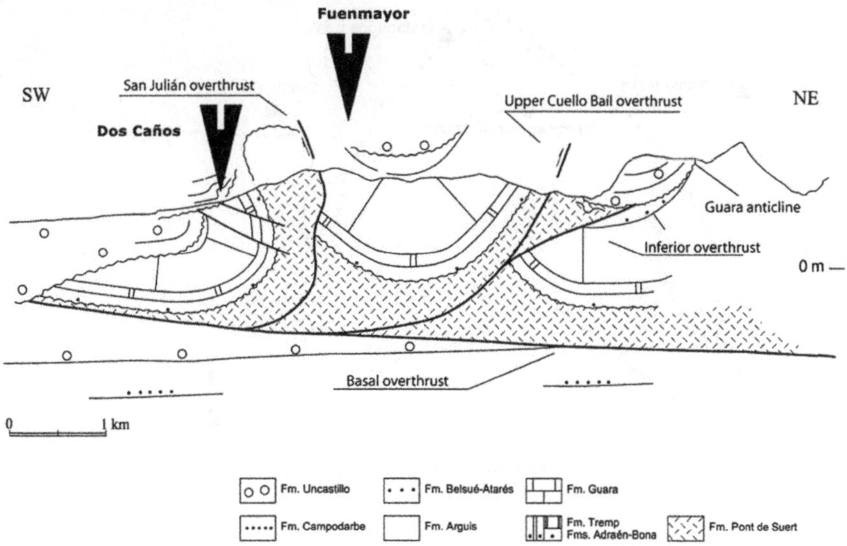

Fig. 4.4 Geological scheme of the study area, according to Millán [20]

However, a pumping test carried out in the only well tapping Dos Caños in 2005 also affected Fuenmayor spring, indicating a hydrological connection, suggested by Cuchí and Villarreol [10]. These authors proposed two hypotheses about interconnection between both aquifers. First one, there are approach points to a shallow underground connection of both aquifers. Another hypothesis establishes a much deeper connection that separates two bodies of different salinity by an interface [10].

According to Cuchí et al. [7], the recharge area shows clear exokarst features. *Lapiaz* structures, mostly *hohlkarren* type [32], are formed by disolution of the surface of the limestone. No karst caves have been discovered in the area. A large part of the recharge area is covered with lithosol (like a barren area). The rest shows thin soil. Maximum humidity storage of the soil has been estimated to be 30 mm [7].

4.3.2 Hydrological Overview

Fuenmayor spring already was referred by Mallada [17]. The behavior of the aquifer is known by the evolution of its spring discharge. From early times, it is known that Fuenmayor presents a strong relationship between rainfall and discharge, with short floods after raining and long low discharge periods during the summers. For this reason, the karst nature of the aquifer was recognized by Trilla and Pascual [33]. Consequently, several actions have been performed in order to regulate the spring, as the building of a water mine suggested in 1955. Later, other actions, as those summarized by Cuchí et al. [7], were made at Fuenmayor and surroundings, in order

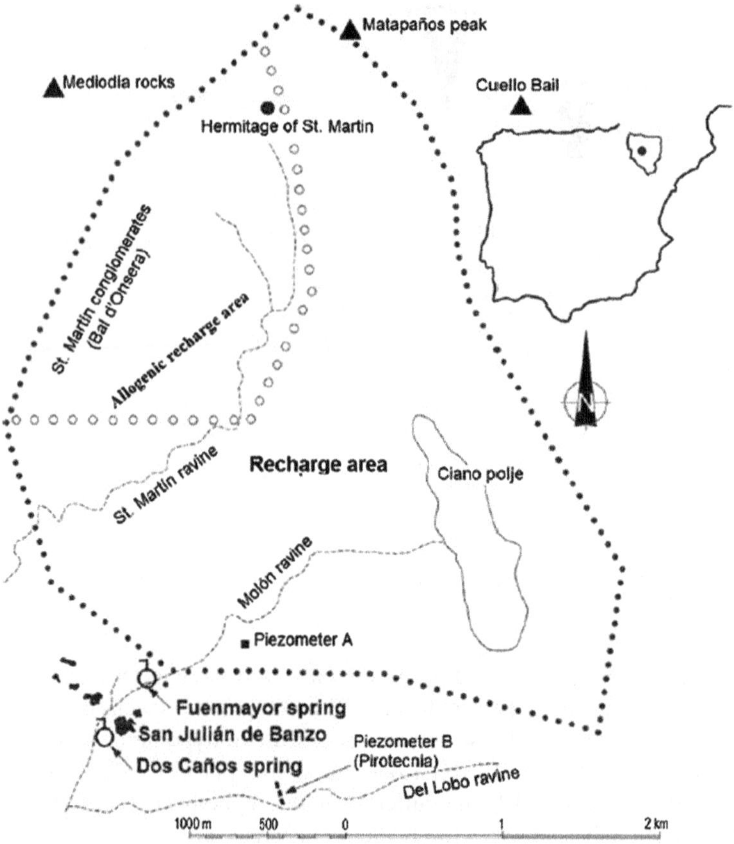

Fig. 4.5 Schema of two structural hydrogeological units: Fuenmayor (*center*), with its hypothetical basin limits, and Dos Caños (*south*) [10] on geological section

to increase the water supply to the city of Huesca. For this purpose, a unsuccessful borehole was drilled at the spring in 1984. Because of this failure, at 1990, two manual water gauges were built by the IGME (Geological Survey of Spain) and CHE (Water Authority of Ebro Watershed) in order to obtain more information on the spring behavior. One was built at the Huesca aqueduct and the other at the spillway of spring.

At 2000, the GTE, Grupo de Tecnologías en Entornos hostiles (Group of Technologies in Hostile Environments), of the Zaragoza University designed and implemented an automatic monitoring station to measure continuously the following parameters: discharge, rainfall, electrical conductivity, water and air temperature. The device was working uninterruptedly until the summer of 2005, when a pumping test from one, until that moment, unused well, dried the spring. A literature review of work carried out on Fuenmayor spring is collected in Table 4.1.

Table 4.1 Literature review of studies about Fuenmayor spring

Date	Fact to be mentioned	Citation
From ancient times	From ancient times, the water from Fuenmayor is used for irrigation and water supply to the local population	Cuchí et al. [5]
1878	Description of this spring in the publication about the geology of Huesca province	Mallada [17]
19th century	Construction of the water conduct to the city of Huesca as an alternative to the use of springs of Ibón and Ángel that capture waters of Huesca watershed	Cuchí et al. [5]
1955	The first one who estimated the water balance of Fuenmayor spring and proposed the construction of a drainage gallery	Lasierra [15]
1974	They explicitly recognize the karst nature of the aquifer, initiate systematic appraisals, correlate discharge with rain and analyze the flow depletion curve	Trilla and Pascaul [33]
1984	City of Huesca performed by percussion, a survey of 60 m. in the *Gorga Mora*, at the header of the spring but it does not have the expected success	Villaroel and Cuchí [34]
1990	*Confederación Hidrográfica del Ebro* installs two gauging weirs in the output of spring, in the Huesca pipe and the spillway of the spring. Initially measurements are controlled visually	Villaroel and Cuchí [34]
1992	A survey for research was performed at level 787, by direct rotation, in the vicinity of Fuenmayor, and 237 m were tested (report 2911-7-0012)	Carcellar [2]
1994	At about 800 m to the southeastern spring, a pilot drilling was made by percussion, reaching 300 meters deep. This survey captured an aquifer, that drains by the *Dos Canos* spring (report 2911-7-0013)	Carcellar [2]
1995	Publication on the hydrogeological potential of the thrust structures on the southern edge of the Sierra de Guara	Octavio et al. [23]
2000	The research group of University of Zaragoza, GTE, installs a continuous monitoring station of output flow, precipitation, electrical conductivity and temperature of water and atmosphere	Villaroel and Cuchí [34]
2003	Water-rock interaction in the referred area	Monaj [21]
2011	Application of cross wavelet power spectrum to quantify an indication of the similarity or strong relation of rainfall-discharge, independently of the energy amplitude of the signals, and to explain some hydrological cycle phenomena in the karst system	Chinarro [3]
2011	About linear methods employing system engineering techniques for the analysis and identification of a hydrological system, considered as a black box model, which disregards information on the internal structure of the aquifer. Under a linear time invariant hypothesis, the application of the simple correlation of spectral analysis and parametric identification of the transfer function generated some interesting results in the monitored spring	Chinarro et al. [4]

(continued)

Table 4.1 (continued)

Date	Fact to be mentioned	Citation
2011	Hydrogeochemical and isotopic characterization of karst-carbonate aquifer	Oliván et al. [26]
2013	Identification techniques used to study the relationship between rainfall and the discharge of karst aquifers in order to examine some of the possibilities they offer and to address their limitations. The paper applies the reviewed techniques to a time series of the Fuenmayor karst spring, in the southern central Pyrenees. An acceptable linear response is shown. The quality of the different models obtained is evaluated with the Nash-Shutcliffe model efficiency coefficient	Cuchí et al. [6]
2013	Delimitation, assessment of recharge, and aquifer hydrodynamics drained by karst spring Fuenmayor	Oliván [25]

4.3.3 Infrastructure and Facilities

The outflow of Fuenmayor karst system flows, from north to south, in an open stream in about 30 m to reach a diversion dam. It is constructed of concrete, with an embankment of 12 m long, a depth not exceeding 0.5 m and a covering area of 50 m^2. Nearby, a small squared building of 10 m^2 provides sheltering for maintenance tools and monitoring equipment. In the basement, water flows are channeled through a gate to manually regulate the flow toward two possible outputs. The first one is to supply piped fresh water to the city of Huesca, with a Thompson weir to measure the water flow. The other output is a bypass to the natural bed of the spring outflows. Downstream, there is a second rectangular gauge to measure the flow from this channel and water from spillways of the dam.

The hydrological monitoring station was installed in 2000 year, with the aim of measuring, on an hourly basis, discharge, accumulated rainfall, electrical conductivity of the water, and air and water temperature [21]. The main elements of this station are:

- The output discharge is measured simultaneously at two outlets by weirs, with a pressure probe PDCR 130/D. They are pressure transducers with a range of measures from 70 mbar (millibar) to 135 mbar, with a linearity of $\pm 0.1\,\%$ and a total error of $\pm 1.5\,\%$ between $-20\,^{\circ}$C and $+80\,^{\circ}$C. It is an active device that requires a 12 V power supply.
- A rain gauge, model 52202 Young, was attached on the roof of the building. It has an area of 200 cm^2, captures with a resolution of 0.1 mm lever tips, and achieves the accuracy of 2 % in measurement up to 25 mm/hr and 3 % up to 50 mm/hr.
- A conductivity probe, model Global Water WQ301, is placed in the inlet of the Huesca aqueduct, measuring in a range of 0–5,000 μS. An interface for communication with the datalogger (data recorder device) is a classical current loop

Fig. 4.6 Datalogger used in the monitoring station, at Fuenmayor spring

4–20 mA. This device requires a 12 V power that should be supplied by the datalogger battery.

- A temperature probe (PT100) consists of a platinum thermo-resistance to measure a temperature range from 100 to 0 °C, with a tolerance of ±0.15 °C at that temperature range. It is located at the spring outflow channel, at the entrance of the dam.
- Outside building, in a shadow area, there is another thermometer to measure the air temperature.
- Data are stored on a *flash memory* of a Campbell CR10X-2M datalogger (Fig. 4.6). Information is discharged periodically into a laptop PC through a PC208W software and a RS232 communication interface. A complementary GSM mobile with antenna on the roof has been used to send real time measurement, although the telephone coverage is poor.
- The power was provided by a lead battery connected with a solar panel, that should be periodically changed by another full charged.
- All data processing was performed by programs developed over software package MATLAB 2010, from MathWorks Inc., Natick, MA. Also, Java development platform J2SE has been used to create, compile and test algorithms for analysis and modeling.

A second weather station was installed in the autumn of 2009, in Ciano Polje, to improve the study of rainfall and atmosphere temperature in the recharge area.

4.4 Analysis of Signals

A first study on the dynamic of the response signals (discharge, water temperature, electrical conductivity of water) in relation of the input signals (rainfall, atmospheric temperature), has been qualitatively studied by Cuchí et al. [8]. The relationship rainfall-discharge shows three classes of behavior. The normal one is a typical hydrograph, with a delay of about 7 h from the start of the rainfall. The rainfall has a double effect on the electrical conductivity. During a few hours, after the start of the increase of the discharge, the EC drops about 10 % of the initial value. After that phase, the parameter increases about 20 μ S/cm. above the pre-rain values. That effect is lengthy and may be detected several weeks after the rainfall. The temperature response of the spring depends on the two studied inputs of the system. The variation of the atmospheric temperature leads to a direct change of the water temperature, with a delay of 3 h. The annual interval of variation of the water temperature, approximately 2 °C, is lesser than the atmospheric one. Moreover, the major peaks of the discharge in relation to heavy rains sheds only slight drops of water temperature.

4.4.1 Sampling Period

The present work only studies the relationship between effective rainfall and discharge. Figure 4.7 represents both time series composed of 25,695 hourly samples—discontinuous for rainfall—recorded for nearly three years, from July 2002 to June 2005, with the exception of a period of pumping test in a nearby well that temporarily affected to the spring outlets.

A very important issue is the sampling period selection. A sampling of one hour was chosen to facilitate the detection of seven hours delay between rainfall and raising of the discharge—which was confirmed by the opinion of local population—plus, the analysis of the hydrographs of isolated rainfall events shows that the raising limb lasts not less than 12 h. The suitability of this assumption was confirmed by the results in the Fuenmayor spring analysis. Moreover, it is generally accepted that sampling should be done eight or ten times during the rise time transient response [24]. In case of choosing a larger sampling period, it would have meant the loss of information about the system dynamics in high frequency. This observation disagrees with frequent cases in the scientific literature, where karst hydrology are performed with larger sampling periods (e.g. 12 h or 1 day).

4.4.2 Time Series Preparation

In the present thesis, evapotranspiration has been discounted from the total rainfall to have the effective rainfall, which was estimated as explained in Sect. 3.2.4.1.

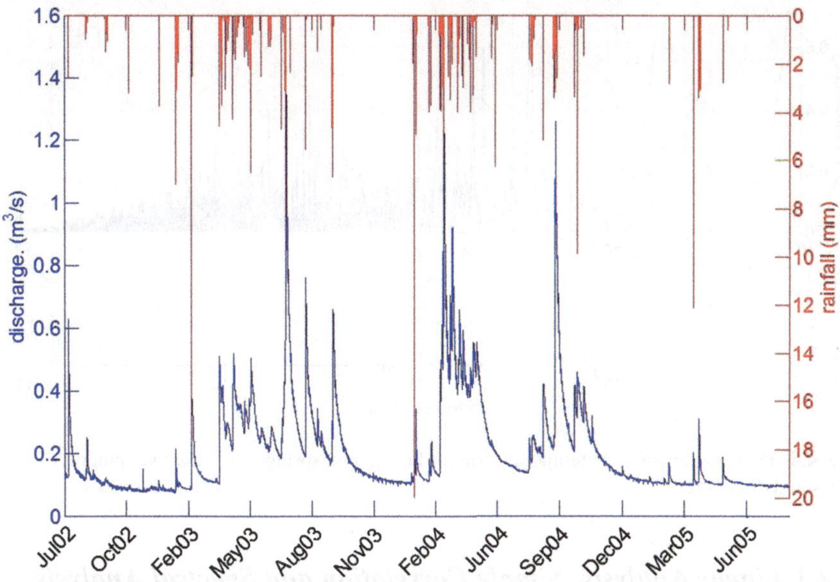

Fig. 4.7 Signals used for the analysis of the Fuenmayor spring: (*red*) effective rainfall; (*blue*) measured discharge

The Eagleman expression (Eq. 3.4) provides a proper estimation of evapotranspiration, considering the relative humidity and the air temperature. Humidity data, required in the formula, was taken from the nearest meteorological observatory, located in Monflorite Airport (16 km south).

The representation of rainfall time series can be seen in Fig. 4.7 above the discharge time series. By direct observation, rainfall shows dry and wet periods including an episode of very severe drought characterized by a long period with low rainfall. For this reason, it can be considered that nearly all system modes have been excited. Hence, if the system modes are excited, these modes can be calculated using system identication techniques. The hydrograph shows short flow peaks related to rainfall events and long flow periods between rainfall seasons. By examining the hydrograph analysis of isolated rainfall events (Fig. 4.7), the growing ramp has a lasting shorter than 12 h, according with the local population observations.

Time series collected from the spring sensors, due to accidental malfunction of the instruments or other factors, can supply raw data with some questionable values, that can distort the later analysis. A new test to detect and correct outliers proposed in (Sect. 2.4.2.4) has been applied to raw data of Fuenmayor. The result is that outliers have been replaced by regenerated values, according to an algorithm of prediction, trending and smoothing.

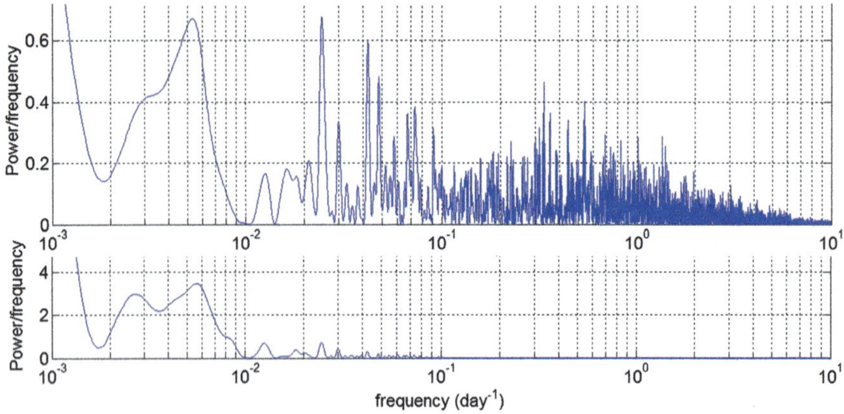

Fig. 4.8 Periodograms for the temporal series of Fuenmayor spring: (*up*) effective rainfall; (*down*) measured discharge

4.4.3 Linear Analysis. Simple Correlation and Spectral Analysis

The input and output series are analyzed by two periodograms as explained in Sect. 2.3. Periodograms (Sect. 2.3.2) of the effective rainfall and discharge of Fuenmayor, shown on Fig. 4.8, are obtained by plotting the frequency values in relation to the power/frequency rate (Eq. 2.5).

From the frequential point of view, the effective rainfall has a main peak at a frequency of 0.0053 days^{-1}. It is a period of 188 days ($\approx$0.5 years) related with the local Mediterranean weather with rainfall seasons every six months. The annual cycle appears as a shoulder at the left of the main peak. The plot also shows secondary peaks at 40 and 23 days without easy explanation. An important conclusion from the frequential analysis is that the effective rainfall is obviously far from white noise.

The discharge exhibits a first peak at 367 days related to the yearly hydrologic cycle, and a second peak, at 188 days ($\approx$0.5 years), corresponding to the seasonal behavior of the local weather. The rest of the higher frequency peaks that appear in the rainfall are highly attenuated in the discharge signal. Thus, the Fuenmayor system behaves as low pass filter.

Figure 4.9 presents the autocorrelation (Eq. 2.4) and the spectral density of the discharge of the Fuenmayor spring. Autocorrelation shows a memory effect of 52 days and correlation factor of $r_k \leq 0.2$). The spectral density function (Eq. 2.8) presents a cut-off frequency of 0.08days^{-1} and establishes a regulation time of 44.2 days. Signals with a period of less than 12.5 days are heavily filtered by the system. These calculated parameters classify Fuenmayor as a Fontestorbes type, in the classification of Mangin [19] expanded by Garay [13].

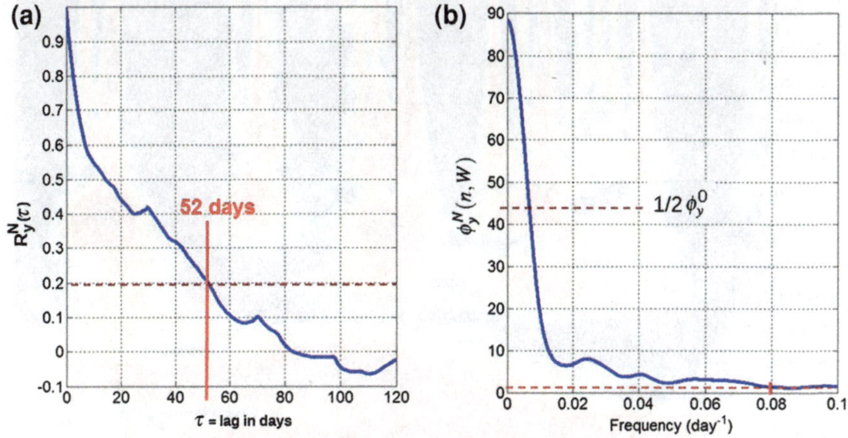

Fig. 4.9 Discharge autocorrelation and spectrum for Fuenmayor spring: **a** Autocorrelation; **b** Spectral density

4.4.4 Wavelet Analysis

The continuous wavelet transform is very suitable to investigate hydrological time series, because of the wide range of possible dominant frequencies, and its efficiency for analyzing non-stationary signals. All wavelet analysis functionalities have been implemented by the authors using the programming languages Matlab 7.0, from Mathworks Inc., and Java tools from Oracle Corporation. Algorithms to code COI, significance levels and phase arrows have been taken from Grinsted et al. [14].

4.4.4.1 Wavelet Power Spectrum of Rainfall Signal

The wavelet power spectra (scalogram) for rainfall (Fig. 4.10, up) assesses the spectral power variance of the rainfall at each scale and at each time lag. Small scale values in the Y-axis (corresponding to high frequency) are represented near the top of the plot, and the large-scale objects (low frequency) are placed toward the bottom.

Figure 4.10 shows continuous wavelet power spectra of rainfall time series of Fuenmayor, applying the expression Eq. 2.13. The greatest number of precipitation events come from the Spring and Autumn periods with a minimum in the Summer. This is confirmed a semiannual oscillation on 180 days scale. About the scale of 4000 days, there are a clear periodical rainy events, with an outstanding spectrum, but is outside the influence cone.

The wavelet mother, according with Eq. 2.9, for the wavelet transform is a bandpass filter for the rainfall signal. To display the scalogram (Fig. 4.10) sampled by band filters, scales have been conveniently written as fractional power of two.

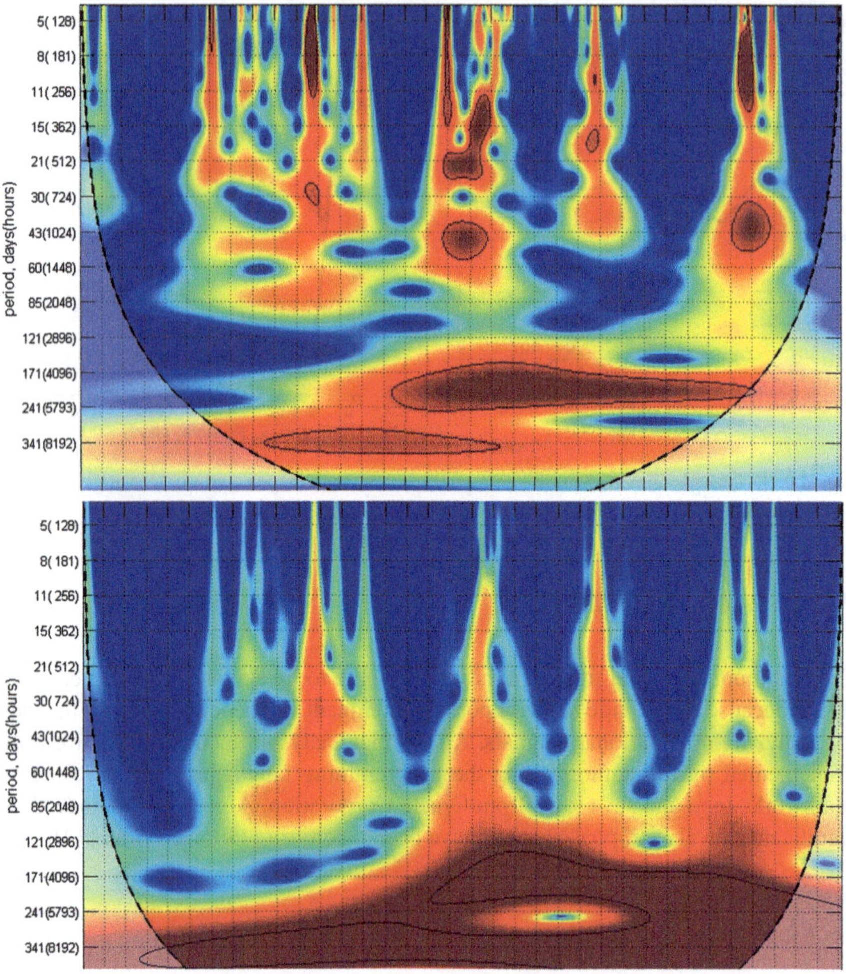

Fig. 4.10 *Up* continuous wavelet power spectrum of rainfall in Fuenmayor basin. *Down* continuous wavelet power spectrum of discharge time series of the Fuenmayor spring. The *y* axis is the wavelet scale (hours). The *x* axis is time (months). *Curved lines* on either side indicate the cone of influence where edge effects become important. The *thick black line* indicates significance level (established at 5 %) for power spectrum. Spectral strength is shown by colors ranging from *deep blue* (weak) to *deep red* (strong). The *thick black* contour designates the 5 % significance level against red noise. The *dashed line* is the cone of influence. The standardized rainfall series has an AR1 coefficient of 0.02

In non-orthogonal analysis, the set of used scales can be chosen arbitrarily. By visual inspection of Fig. 4.10, some features are perceptible:

- The spectral map shows a semiannual cycle following the typical pattern in the Mediterranean climate, which is observed by the continuous dark red in horizontal direction, in the scales range of 171–241.
- Another highlight spot is located beyond the scale 341, closer to the scale 360, in all timeline values, which corresponds to the annual hydrological cycle.
- The black dashed line represents the COI, to highlight the influence of the ends of the range of data. Since COI is a trust boundary, findings outside the cone would be under suspicion of unreliability.
- In rainy seasons (6–13, 16–20, 23–25 month intervals), the rain reveals a richer spectrum with frequency components in the 200–2000 h band (about 8–80 days). In dry intervals, these components are absent.
- The plot shows rainfall events as small isolated zones in the high frequency band that requires to zoom the spectral picture.

4.4.4.2 Wavelet Power Spectrum of Discharge Signal

Analogously, at a glance on Fig. 4.10 down, the power spectrum of the output spring reveals also natural oscillations of discharge regime. The distribution of energy in the discharge signal and the concentration in certain areas can be observed by different color gradient levels. The hue gives a quantitative idea of the frequency contained in a signal at a certain time, so it represents the possible value range of frequencies contained in the discharge signal. Dark red peaks of the discharge emphasize visibly the annual and semiannual cycles.

The wavelet window, acting as a filter, can be moved both along the time line and along the scales with a defined granulation. So, more details about a specific area of the spectral map can be obtained. The upper spectrum is zoomed, to observe, in high frequency bands, details on possible cycles of the discharge in short periods. Similarly, in case of an extended series of data, the filter can be shifted towards lower frequency bands (upper scales), to find likely rainy cycles and drought cycles along lengthy periods of time (Fig. 4.11). The longer the data series is, the less COI will influence the ends, and the more accuracy of correlation will be achieved, especially at low frequencies. It is interesting to know is that the wavelet transform can depict and place the assessment of the frequency component at the time it occurs. However, Fourier procedures can reveal that there are some certain frequencies in the signal, but not when they happen.

By a simple visual examination of the Fig. 4.11, the wavelet spectral power exhibits a quasiperiodic behavior in the high frequency. The discharge power spectrum discloses the weight of each frequency carried by the signal. Between the dark blue areas that represent the common dry season of each year, there are a few spectral power peaks in the frequency range approximately between 30 and 43 days. This happens every year, although with slight differences of hue. These results are similar to those obtained by linear methods.

The plot also shows erratic rainfall events forming small isolated zones in the very high frequency band. Obviously, dry intervals are characterized by the scarceness of

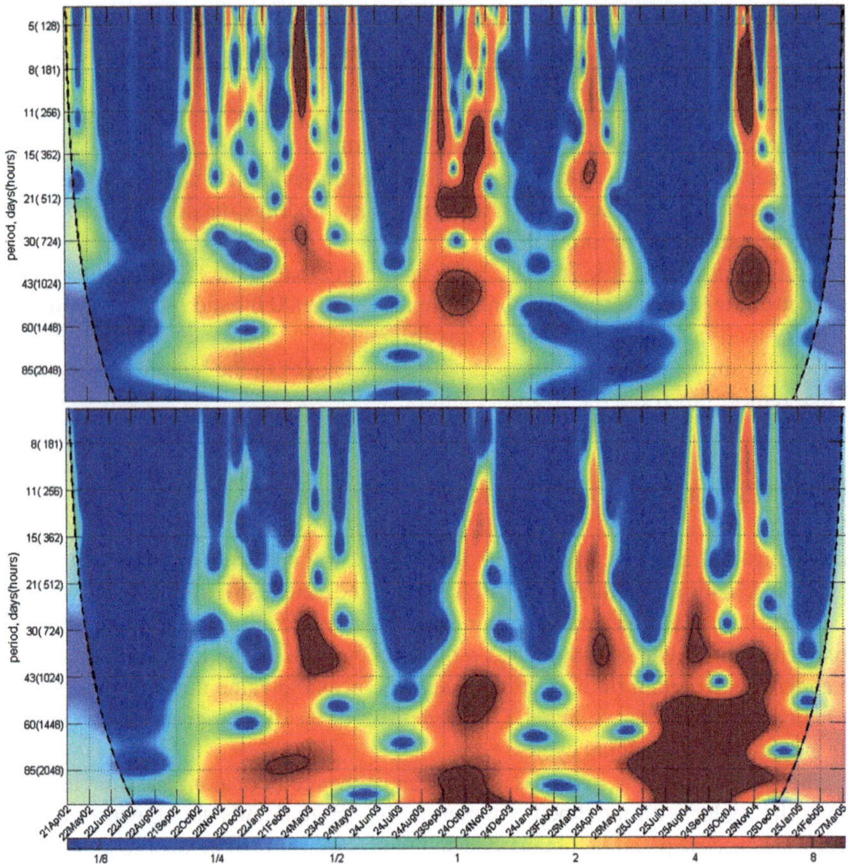

Fig. 4.11 Wavelet spectral power of the rainfall (*up*) and the discharge (*down*) in Fuenmayor spring zoomed to display details in high frequency bands

spectral components. Because of the rainfall effects over discharge are attenuated in wet period, the karst groundwater system behaves as a loss-pass filter. Importantly, in the discharge spectrum graph, the weaker color in certain areas, outside the COI bounds, indicates the area of unreliable values.

Therefore, the resolution advantage of the wavelet transform permits detection of changes in nonstationary of discharge and rainfall, with exceptional localization both in time and frequency domains.

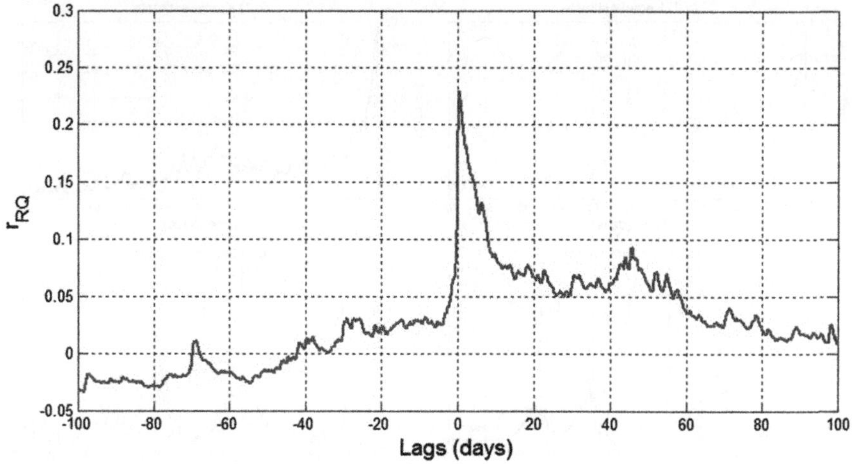

Fig. 4.12 cross-correlation between effective rainfall and discharge in Fuenmayor spring

4.5 System Identification

The key problem in the system identification is to find a model structure, with sufficient flexibility to suit the karst dynamics. Once the signals observed in Fuenmayor spring have been analyzed, in the following paragraphs, the relationship between precipitation and discharge signals of Fuenmayor spring are studied following the different techniques reviewed in Chap. 2.

4.5.1 Nonparametric System Identification

This section applies the main nonparametric system identification methods, given in Sect. 2.7, to the effective rainfall and discharge in Fuenmayor spring.

4.5.1.1 Cross-Correlation. Impulse Response

The cross-correlation coefficient between the effective discharge and discharge of the Fuenmayor spring is depicted in Fig. 4.12, according to Eq. 2.34 from Sect. 2.7.1.1. An estimate of the unit hydrogram with a sharp pointed peak at 14 h suggests a fast flow into a highly karstified system. The recession limb can be divided into two sections. The first one, from 14 h to approximately 10 days, has a fast decay. In the second, the decay is slower and more extended in time. This suggests that the behavior can be identified by a system of second order.

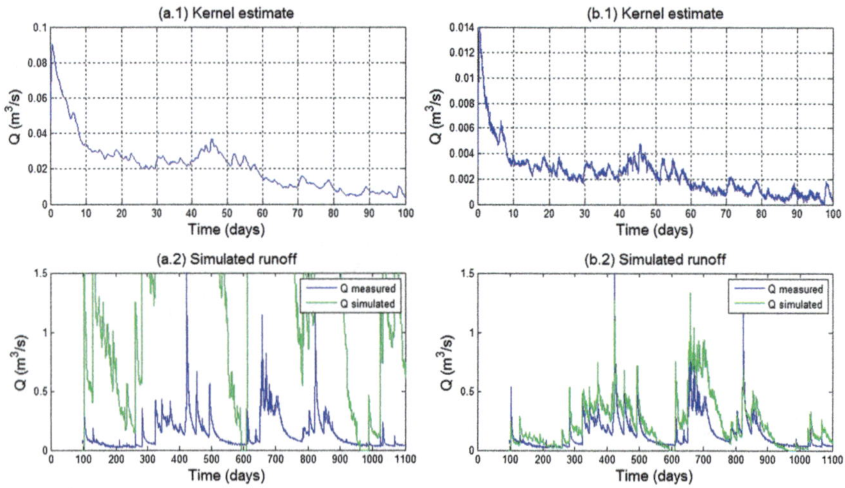

Fig. 4.13 Kernel estimation of Fuenmayor spring based on cross-correlation: **a** without prewhitening and **b** with prewhitening

Using theory described in Sect. 2.7.1.1 (Eq. 2.35), the cross-correlation coefficient to estimate the impulse response is represented in Fig. 4.13a.1, and, applying the convolution sum, it is possible to simulate the system. Figure 4.13a.2 shows the result of this simulation for the effective rainfall in comparison with measured discharge. As can be seen, the discharge is greatly overestimated. This model based on the cross-correlation coefficient gives an efficiency factor of $E = -314.14$ for the Fuenmayor data applying Eq. 2.51, which gives a poor efficiency value, according discussed in [6]. However, as has been stated in Sect. 3.2.4.1, the effective rainfall of the Fuenmayor karst system is far from being white noise. To overcome this problem, the whitening filter technique has been applied. A whitening algorithm performs a decorrelation that transforms a set of random variables having a known covariance matrix into a set of new random variables whose covariance is the identity matrix. Figure 4.13b.1 shows the impulsional response estimate obtained after the whitening. Figure 4.13b.2 shows that in this case the simulated discharge fits much better than the previous one. The simulation model presents an efficiency value of $E = 0.095$. Finally, the estimated impulsional response is near zero after 100 days.

Figure 4.14 shows the linear coherence function. The Fuenmayor system is highly linear for low frequencies up to 0.08 days^{-1} (signals with period greater than 12.5 days). For higher frequencies, the linearity decreases slowly. Compared with others springs analyzed in the literature, Fuenmayor presents a good linearity, similar to Torcal of Antequera [27], and in contrast with Torremolinos [1], Simat [18], Sant Josep [12] and Ancón [30].

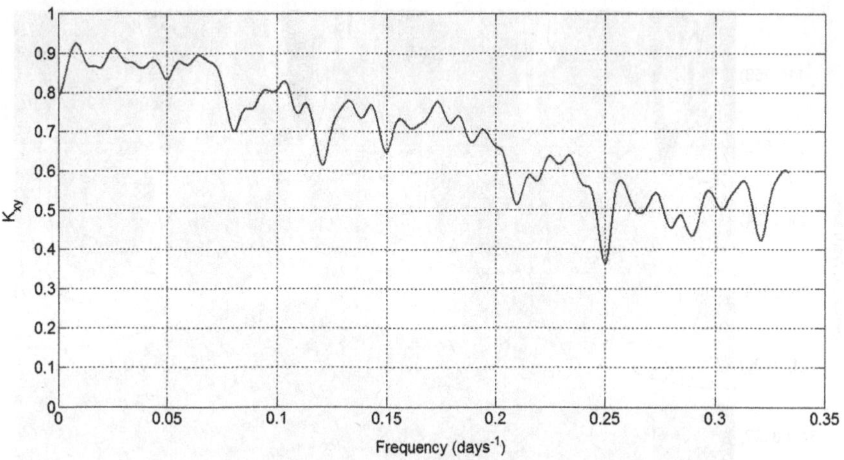

Fig. 4.14 Linear coherence function of Fuenmayor spring

4.5.1.2 Wavelet Coherence

Figure 4.15 shows the wavelet coherence (Sect. 2.7.2.3) between the rainfall and the discharge at Fuenmayor spring, following the expression (2.47). In each point of the frequency-time map, the coherence result is a value between 0 for dark blue color, and 1 for dark red color. A value 1 for a given frequency band, indicates that the response energy is 100 %, due to the input rainfall signal enters in a linear relationship. The 5 % of significance levels are plotted with black thick line. The left axis is the wavelet scale (hours). The bottom axis is time. Curved lines on either side indicate the COI. The spectral strength spreads out from weak values (blue) to strong ones (dark red).

As expected, the wavelet coherence (Sect. 2.7.2.3) shows wide areas of a strong correlation between the input and output signals, with values close to 1, after applying the Eq. 2.47. However, the spectral coherence presents several blue areas (i.e. with very low coherence) in some areas, inside the COI. This suggests there are two behaviors of the system:

- Strong coherent behavior that happens when the system clearly reacts in consequence of the rain.
- A weak or non-coherent behavior that occurs when it rains after a dry season (0–6,15–18, 28–30 months).

There are periods in the range of June–October 2002, June–October 2003 and May–October 2004, in which a non-coherent behavior occurs, showing the non-linearity of the system. Identified empty intervals or dark blue areas correspond to dry seasons. When it rains after a dry season, the soil and epikarst zones need to be wetted, and its maximum water storage replenished before the effective rainfall has some influence on the groundwater and consequently on the discharge output.

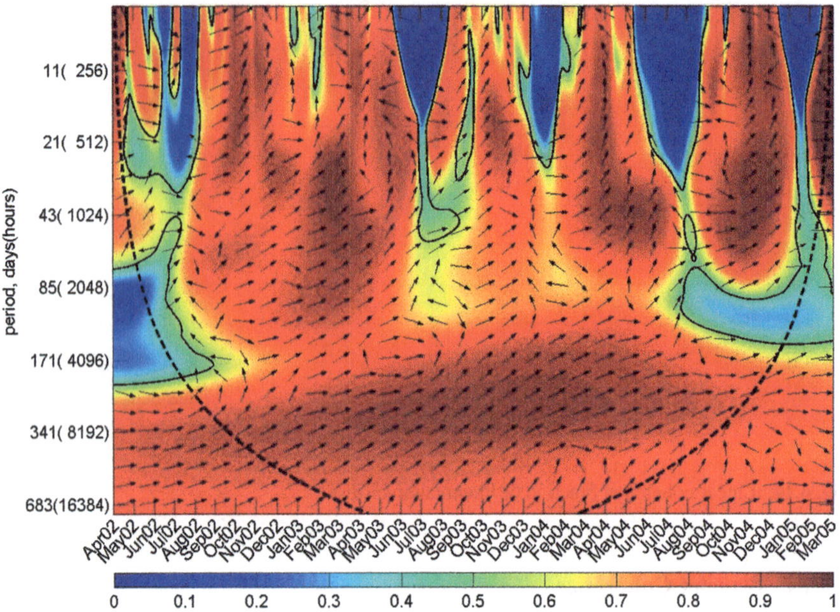

Fig. 4.15 Coherence spectrum between rainfall and discharge in Fuenmayor spring. The *arrows* represent the phase angles between the effective rainfall and the discharge in the respective frequency band

In wet periods, the linear response of the system and high value of coherence can be easily observed. Conversely, when the system is dry, the reaction to the rain event is not linear. At this point, wavelets can provide their essential support. Thus, the wavelet coherence is useful to determine in which period of time it is recommendable to apply linear methods and when to consider the system as a non-linear model to be analyzed with different non-linear tools. In conclusion, two main modes have been identified in the system: the wet mode with a linear behavior and the dry mode with non linear behavior. This variation of the coherence evidences the nonlinearities of the system.

4.5.2 Parametric System Identification

4.5.2.1 Identification of the Impulse Response

The next step is the identification of the impulse response of the system. Two of the methods presented in Sect. 2.7.1.2 have been applied in Fuenmayor: the Wiener Hopf summation equation (Eq. 2.36) and the Least Squares method (Eq. 2.39). As the estimation is performed by cross correlation, where the estimated impulsional

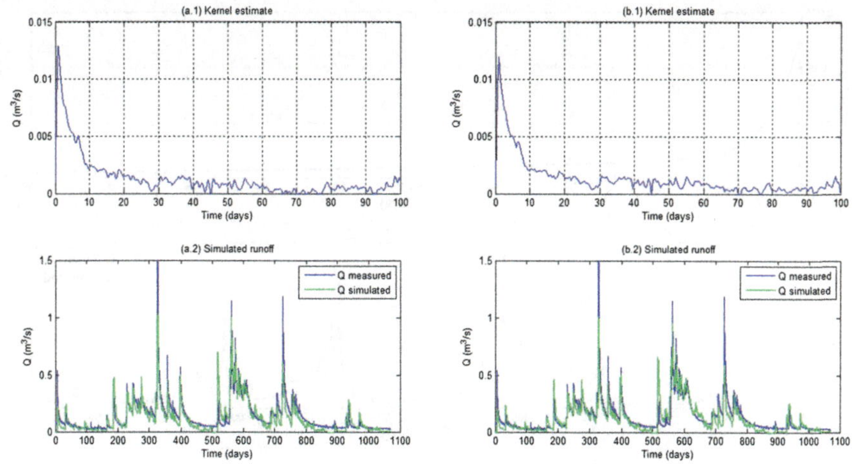

Fig. 4.16 Kernel estimation of Fuenmayor karst system based on: **a** Wiener-Hopf summation equation and **b** error minimization

response is near zero after 100 days, only the first 2,400 samples (100 days) of the kernel or impulse response have been calculated. It should be stated that both methods presented computational problems with excessive execution times (several hours in an Intel Quad Core at 2.4 GHz) and high memory demands (with 2 GB of physical memory the application filed out of memory). Thus, a decimation process was applied to convert the hourly series into another series with a sample for every 12 h. After the decimation the calculus was possible. To compare the simulated with the measured discharge, an interpolation process was applied.

Figure 4.16 shows the estimated kernels and the simulated discharge for both methods. The model obtained by the Wiener-Hopf summation equation presents an efficiency (Sect. 2.8.2) of $E = 0.7941$, calculated by Eq. 2.51. The model obtained by error minimization shows $E = 0.7934$. Thus, both methods are very similar and reproduce the Fuenmayor behavior quite well.

4.5.2.2 Identification of Transfer Functions

Applying to the Fuenmayor spring some parametric identification methods, introduced in Sect. 2.8, and proposed by Ljung [16] for linear system, such as ARX, OE, ARMAX and BJ, several model structures has been calculated with different orders. The model with the best efficiency (Eq. 2.51) is an OE model with two poles, one zero and seven delays. Applying Eq. 2.25 the following transfer function is obtained:

$$\frac{Q_s(z)}{Q_u(z)} = \frac{0.01289 - 0.01286z^{-1}}{1 - 1.9882z^{-1} + 0.9882z^{-2}} z^{-7} \tag{4.1}$$

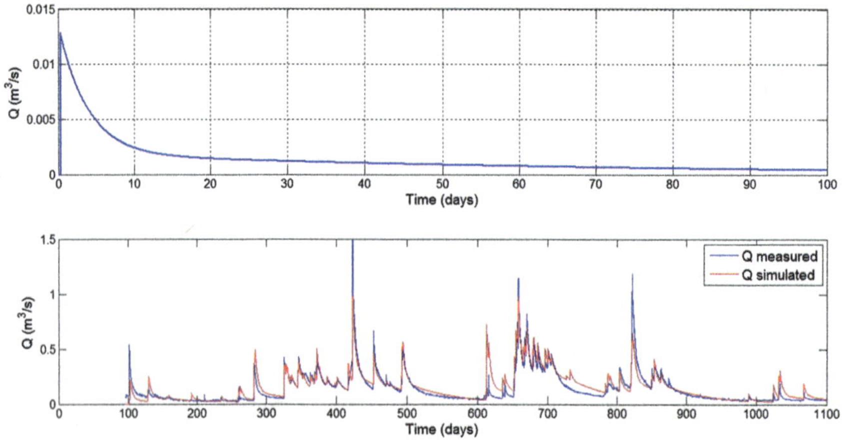

Fig. 4.17 Impulse response (**a**) and simulated runoff (**b**), in the parametric identification of Fuenmayor spring

As explained in Sect. 2.5.1, poles and zeros are complex quantities that determine the dynamic behavior of the aquifer system. In this case, the TF poles are z1 = 0.9994 and z2 = 0.9888 corresponding to recession coefficients $\alpha_1 = 0.014$ days^{-1} and $\alpha_2 = 0.2712$ days^{-1}. The first pole value (the slow pole) is similar to the last depletion coefficient, calculated by Trilla and Pascual [33] using the Maillet method for drought periods. The function also has seven delays (seven units of time). That is the observed lag, around 7 h, in the response of Fuenmayor to the rainfall.

Figure 4.17 shows the Transfer Function impulse response and the relationship between measured and simulated discharges. According to expression of Nash and Sutcliffe [22] (Eq. 2.51), the coefficient E evaluate the quality of the simulations obtained for this new model, E = 0.8164. This is the best efficiency of all the models tested. Moreover, it is a manageable model with only five parameters that can be used for analysis, simulation and prediction purposes. A standard procedure in system engineering is to analyze the system model through a set of parameters characteristic of the unit step response. For example:

- Static gain = 3.941. Ratio of the system output and the system input values under the steady state condition.
- Rise time = 133 days. The time required for the output to reach 90 % of the steady state final value.
- Overshoot = 0. Ratio of the maximum peak value (minus the steady state final value) and the steady state final value. Fuenmayor spring does not show oscillating behavior because both poles are real.

The use of these kind of parameter may enable a new comparative analysis of the karst system to be carried out in a standard way. The prediction power of the proposed TF function has been validated with new data set for the year 2010.

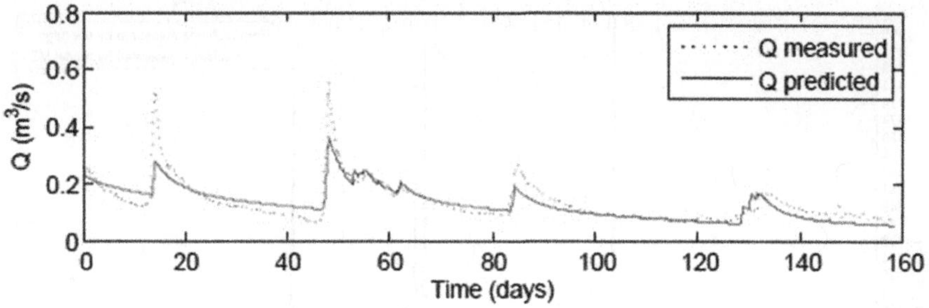

Fig. 4.18 Prediction of discharge in the first half of year 2010, obtained by parametric identification of the transfer function

A model with an efficiency factor of E = 0.8164 may be valid for many applications in karst hydrology. However, in some cases, a deeper study must be performed to obtain more precise models or analyze the discrepancies with the observed behavior. The identification technique, chosen at this part of the thesis, assumes that the system is linear and time stationary.

Figure 4.18 shows the predicted and measured discharge for that period.

4.6 Nonlinear Identification

The prior knowledge on physics of the system, and the previous results from linear identification and spectral analysis, can address the tuning a nonlinear model structure. With one output, i.e. the discharge, and the precipitation as unique input, obtained from Fuenmayor aquifer during 2002–2005 period, three models based on Hammerstein-Wiener have been tried, with different configurations of the component blocks, according with Sect. 2.9.1.2.

For the **first model M1**, the linear block is the same linear transfer function identified in Sect. 4.5.2.2. For the nonlinear blocks, the input and output nonlinearity estimators, two piecewise linear functions with different paremetrs have been tested. The best model among M1 configurations obtained a Nash-Sutcliffe eficiency of E = 0.8174, by Eq. 2.51.

A **second model M2** based on the same structure Hammerstein-Wiener with same linear transfer function identified in Sect. 4.5.2.2. The nonlinear blocks, are a polynomial function in the input of third grade and a piecewise linear function with 10 units. After testing several configurations, the best model with the structure M2 has a Nash-Sutcliffe eficiency of E = 0.8281.

Finally, for the **model M3** an advanced nonlinear model for the Hammerstein-Wiener structure, maintaining the same linear transfer function in Sect. 4.5.2.2. The first block is a wavelet neural network (wavenet) and the second one is a piecewise linear functions with 10 units The running test draws out a Nash-Sutcliffe eficiency of E = 0.9276.

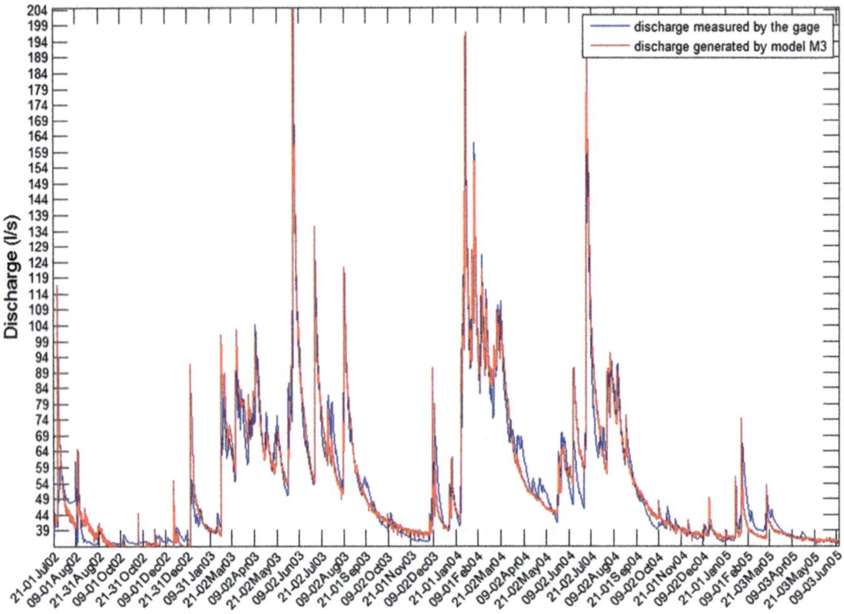

Fig. 4.19 Verification of model M_3 by Hammerstein-Wiener structure for Fuenmayor spring

Therefore, model M3 is the best one obtained for the next stage about prediction, in which a good model is required to forecast the behavior of the spring in a long horizon prediction. Figure 4.19 illustrates graphically the verification of model M3. Moreover, this result leads to those obtained in models of linear identification.

4.7 Conclusion

The Fuenmayor spring has continuously been monitored for identification purposes, to study the behavior and optimize the management of a karst system. Firstly, there is an examination of the model through classical spectral analysis and system identification techniques under a linearity hypothesis.

The Parametric Identification of Fuenmayor has found the first transfer function in this aquifer, proposed under the hypothesis of a linear and time invariant response. A transfer function, reduced to only five parameters (one zero, two poles, the gain and the delay), may describe very properly the hydrologic behavior of this karst system. Moreover, Fuenmayor shows an acceptable linear response.

The use of wavelet transforms allowed the study of the main behavior characteristics of the Fuenmayor spring model, cycles (annual and semiannual) and low-pass filter behavior on the same way that with linear tools. In addition, the cross-wavelet tools revealed the existence of several system operating modes. Moreover, thorough wavelet coherence analysis, the moment when non-linearities appear in the system was detected, which coincides with two main modes identified in the system: the wet mode with a linear behavior and the dry mode with non linear behavior. Therefore, wavelet coherence is an useful tool to reveal certain nonlinearities very well located in drought periods.

Through linear parametric identification methods, a model structure with an efficiency value of 0.8164 (near a perfect model) has been obtained. Considered a linear system, the Fuenmayor spring shows a good approximation between the estimated and observed data. However, the rainfall-discharge model shows nonlinearities revealed by the wavelet coherence spectrum. Moreover, a better characterization of the system has been developed to inquire a deeper knowledge of the system. In this way, some models based on Hammerstein-Wiener structure have reached an eficiency value of 0.9276, that surpasses the results obtained in linear models using the same efficiency criterion.

System Engineering techniques are well developed scientific procedures that can be useful to characterize the hydrological response of the Fuenmayor karstic system.

References

1. Andreo B, Jiménez P, Durán J, Carrasco F, Vadillo I, Mangin A (2006) Climatic and hydrological variations during the last 117–166 years in the south of the Iberian peninsula, from spectral and correlation analyses and continuous wavelets analyses. J Hydrol 324:24–39
2. Carceller MT (2007) Medida de niveles piezométricos y caudales en puntos de la red oficial de control de aguas subterráneas en la cuenca del Ebro. Technical report, Ministerio de Medio Ambiente. Gobierno de España. Secretaría general del territorio y la biodiversidad. Dirección general del agua. Confederacion Hidrográfica del Ebro
3. Chinarro D (2009) Análisis sistémico de manantiales kársticos. Aplicación al manantial de Fuenmayor (San Julián de Banzo, Huesca). Technical report, I3A. Zaragoza University. J.L. Villarroel
4. Chinarro D, Villarroel J, and Cuchí J (2011). Wavelet analysis of Fuenmayor karst spring, San Julián de Banzo, Huesca, Spain. Environmental Earth Sciences, Special Issue:1–13
5. Cuchí JA, Villarroel JL, Manso J (2002) Características del manantial de Fuenmayor, San Julián de Banzo, Huesca. Geogaceta 31:75–78
6. Cuchí, JA, Chinarro D, Villarroel JL (2013) Linear System Techniques Applied to the Fuenmayor Karst Spring, Huesca (Spain). Environmental Earth Sciences, pp 1–12
7. Cuchí J, Villarroel J, Carceller T (2006) Comportamiento del manantial de Fuenmayor, (San Julián de Banzo, Huesca) durante la sequía de 2005. In Congreso Internacional sobre: El agua subterránea en los países mediterráneos, AQUAINMED06, pp 24–28
8. Cuchí J, Villarroel J, Carceller T, Azcón A (2005) Comportamiento del manantial de Fuenmayor (San Julián de Banzo) durante la sequía de 2005. Publicaciones del Instituto Geológico y Minero de España. Serie: Hidrogeología y Aguas Subterráneas., 18:135–143
9. Cuchí J (1998) Esquema general de las unidades hidrogeológicas del alto aragón. Lucas Mallada 10:87–105
10. Cuchí J, Villarroel J (2002) Análisis de respuesta de Fuenmayor. Primeros resultados 31:75–78
11. Cuchí J, Setrini J (2004) Primera Aproximación al cálculo de avenidas en los cañones fluviokársticos de la Sierra de Guara. Lucas Mallada 11:94–103

12. Esteller M, Morell I, Antigüedad I and Garay P (1996) Comportamiento del sistema kárstico drenado por el manantial de Sant Josep (Vall d'Uixó, Castellón). In In: Recursos hídricos en regiones kársticas, pp 381–398
13. Garay P (2002) Tipología de acuíferos kársticos basada en el análisis de correlación y espectral de sus descargas naturales: una propuesta de revisión y ampliación de la clasificación de mangin. In Karst and Environment (F. Carrasco and B. Andreo), pp 99–104. Patronato Cueva de Nerja. Nerja
14. Grinsted A, Moore JC, Jevrejeva S (2004) Application of the cross wavelet transform and wavelet coherence to geophysical time series. Nonlinear Proc Geophys 11:561–566
15. Lasierra M (1955) Informe sobre el abastecimiento de aguas a la ciudad de Huesca. Technical report, Ayuntamiento de Huesca
16. Ljung L (1999) System Identification., Theory for the UserPrentice Hall, N J
17. Mallada L (1878) Descripción física y geológica de la provincia de Huesca. Imprenta Manuel Tello, Madrid
18. Mangin A and Pulido A (1983) Aplicación de los análisis de correlación y espectral en el estudio de acuíferos kársticos. Tecniterrae, pp 53–65
19. Mangin A (1982b) L'approche systemique du karst, consequences conceptuelles et methodologiques., Larra, In InReunión monográfica sobre el karst, Larra
20. Millán H (1996) Estructura y cinemática del frente de cabalgamiento subpirenaico en las sierras exteriores aragonesas. Tesis Doctoral. PhD thesis, Universidad de Zaragoza
21. Monaj D (2003) Interacción agua-roca en casos concretos del prepirineo altoaragonés. Technical report, Universidad de Zaragoza, Grade final project for Chemical Engineer
22. Nash J, Sutcliffe J (1970) River flow forecasting through conceptual models. Part I. A discussion of principles. J Hydrol 10:282–290
23. Octavio F, Ortiga M, Román JS (1995) Posibilidades hidrogeológicas de las estructuras cabalgantes del borde meridional de la Sierra de Guara. In VI Simposio de Hidrogeología, volume XXI, pp 57–64.
24. Ogata K (1987) Discrete-Time Control Systems. Prentice-Hall International Editions
25. Oliván C (2013) Delimitación, evaluación de la recarga y funcionamiento hidrodinámico del acuífero drenado por el manantial kárstico de Fuenmayor (prepirineo aragonés). PhD thesis, University of Zaragoza
26. Oliván C, Lambán L, Cuchí J, Villarroel J (2011) Caracterización hidrogeoquímica e isotópica del acuífero carbonatado-kárstico drenado por el manantial de Fuenmayor (Huesca). In Cuatro décadas de investigación y formación en aguas subterráneas. Libro homenaje al Profesor Emilio Custodio, volume 1, pp 393–402. Asociación Internacional de Hidrogeólogos (Grupo Español), Zaragoza
27. Padilla A, Pulido A (1995) Study of hydrographs of karstic aquifers by means of correlation and cross-spectral analysis. J Hydrol 168:73–89
28. Pinilla L (1996) Estudios para la delimitación del perímetro de protección del acuífero del manantial de san Julián de Banzo (Huesca). Confederación Hidrográfica del Ebro 1:58
29. Puyal M, Cuchí J, Manso J (1998) Hidroquímicas de aguas naturales del pirineo altoaragonés. Lucas Mallada, pp 173–187
30. Rodríguez J, Antigüedad I, Llanos H, TTM (1994) Caracterización hidrodinámica de los acuíferos kársticos del Pan de Guajaibón, Pinar del Rio, Cuba, a partir de las respuestas naturales de sus surgencias. Universidad de Granada, In El Karst y los Acuíferos Kársticos, pp 137–166
31. Rodriguez-Vidal J (1986) Geomorfología de las sierras exteriores oscenses y su piedemonte, vol 4. Colección de Estudios Altoaragoneses, Huesca
32. Sweeting M (1973) Karst Landforms., Selected GlossaryColumbia University Press, New York
33. Trilla J, Pascual I (1974) Análisis de hidrogramas de una surgencia cárstica (Fuenmayor). Centro de Estudios, Investigación y Aplicaciones del Agua 87:20–28
34. Villarroel JL, Cuchí JA (2004) Estudio cualitativo de la respuesta, de mayo de 2002 a abril 2003, del manantial kárstico de Fuenmayor (San Julian de Banzo, Huesca) a la lluvia y a la temperatura atmosférica. Boletín Geológico y Minero 115:237–246

Chapter 5
Analysis and Identification of Collins Glacier Drainage

> *... navegado bajo el Almirante don Gabriel de Castilla con tres barcos a lo largo de las costas de Chile hacia Valparaiso, i desde allí hacia el estrecho, en el año de 1603; i estuvo en marzo en los 64 grados i allí tuvieron mucha nieve. En el siguiente mes de abril regresaron de nuevo a las costas de Chile.*
> Dutchman sailor Laurenz Claesz's testimony about Gabriel de Castilla's journey to Antarctic (writen probably by 1607)

5.1 Introduction

Glacier watershed discharge is part of an intricate process as explained in Sect. 3.3.3. In the Collin glacier, the annual cycle is divided between a cold inert period, with snow accumulation, and a warmer active period. Between both, there is a transition season with complex melting processes, hardly any water flow, and multiple freezing and refreezing operations, since ice snowpack interface follows the air temperature changes.

A number of preferential flow paths are created and/or reactivated inside the glacier, as *bédières*, moulins crevasses and horizontal conduits. The sudden opening of conduits may produce outbursts, which cause irregularities in the glacier discharge in the form of sharp peaks in the output signal. The active period is considered the target of this system identification study.

The chapter focuses on applying wavelet spectral analysis and system identification techniques to study two large time series, from 2001 to 2011, of proglacial water discharge from Collins glacier (King George Island, Antarctica) and air temperature from a nearby meteorological station. Through the study of the dynamic behavior, correlating both parameters, a suitable mathematical model can be found. In addition, the active periods of the glacier have analytically been determined, founded on the spectral coherence.

D. Chinarro, *System Engineering Applied to Fuenmayor Karst Aquifer (San Julián de Banzo, Huesca) and Collins Glacier (King George Island, Antarctica)*, Springer Theses, DOI 10.1007/978-3-319-08858-7_5

The objective of this chapter

The purpose of this chapter in the first stage, after determining the active period boundaries in each annual cycle, is to use system engineering methods and wavelet analysis to study the time series of discharge in the active period and air temperature, and to reveal the correlation between them. Also, some seasonal patterns of the glacier are highlighted, with valuable information emerging from the application of linear and nonlinear techniques from the system identification.

Outline of this chapter

Section 5.2 provides a **geographical framework** of the area under study in Antarctica, specially focused on the climate (5.2.2) and features (5.2.1) of King George Island where Collins is situated. A description focused on **Collins glacier** is in Sect. 5.3, with data pertaining to its size and evolution, along with a geological overview, hydrological explanation of the catchment area (5.3.1), and a description of sensoring instruments and facilities (5.3.2) used in retrieving data. Additionally, some processes applied for discharge volume measurement (5.3.2.1) and the procedure for weather measurements (5.3.2.2) are mentioned.

In order to find causality between the weather phenomena and the glacier system, Sect. 5.4, **analysis of time series**, deals with the issues of discharge (5.4.2) and the meteorological variable air temperature (5.4.3), considered to be the major influence on glacier internal ablation and, consequently, the direct responsible for glacier discharge. The Sect. 5.4.5 contains a wavelet power spectrum study of discharge and temperature, showing the cycles in an ample spectrum of frequencies. Air temperature-discharge **coherence** (Sect. 2.7.2.3) reveals the cross cycling behavior between input and output signals in the system, and is necessary to bound the active periods of the glacier. **Seasonal Effective Core** calculus, Sect. 5.4.7, illustrates the process of the estimation SEC by cycles, calculating the limits of the active period separators related to the coherence levels considered.

The aim of Sect. 5.5, **system identification**, is to obtain a mathematical model of the hydrological dynamics of the glacier, after assuming that the temperature is the principal cause of the discharge variability. **Linear model by cycles** (5.5.1) is a black-box parametric identification through a simple model, such as the Output Error (OE) with two poles and one zero. This simplification makes the subsequent explanation and further analysis easier with a minimal loss of accuracy. **Nonlinear model by cycles** (Sect. 5.5.2) improves the quality of the resulting models, with the nonlinear Hammerstein-Wiener model

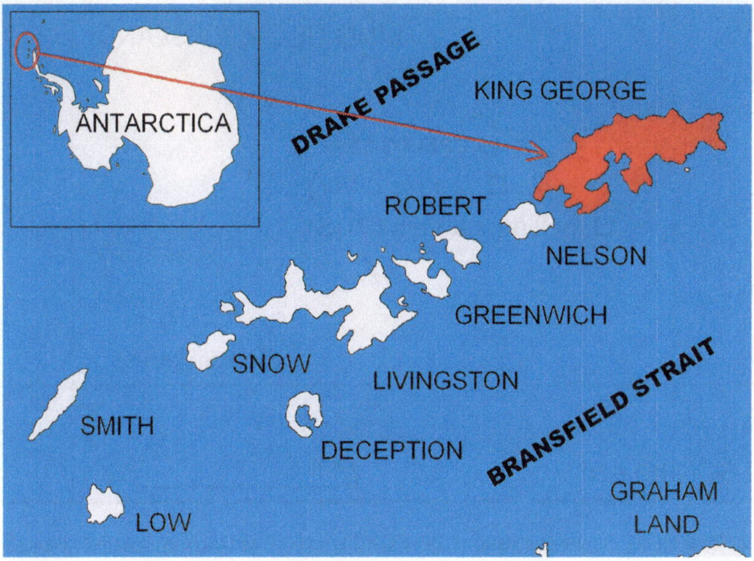

Fig. 5.1 Map of King George, South Shetland archipelago. Antarctic Peninsula. (*Source* Artigas Base, Uruguayan Antarctic Institute)

structure, maintaining OE as the linear block and different structures in the nonlinear blocks. **Glacier Coalescent Model** (GCM) (Sect. 5.5.3) tries to represent a general behavior of the glacier system by blending all seasonal effective cores.

Finally, in the **Conclusion** (Sect. 5.6), some important assertions related to linear and non-linear identification of Collins glacier are presented, as are the main results.

5.2 Geographical Framework

The South Shetland archipelago is a group of islands located on the west side of the Atarctic Peninsula (62 °S 58 °W), separated from Peninsula by Brasfield strait. The archipelago comprises: King George, Livingston, Greenwich, Robert, Nelson islands and others (Fig. 5.1), some of them with scientific bases, as the Spanish Juan Carlos I base in Livingston. The climate in this subpolar region presents variations on maritime and atmosphere conditions; thus the zone is generally considered highly

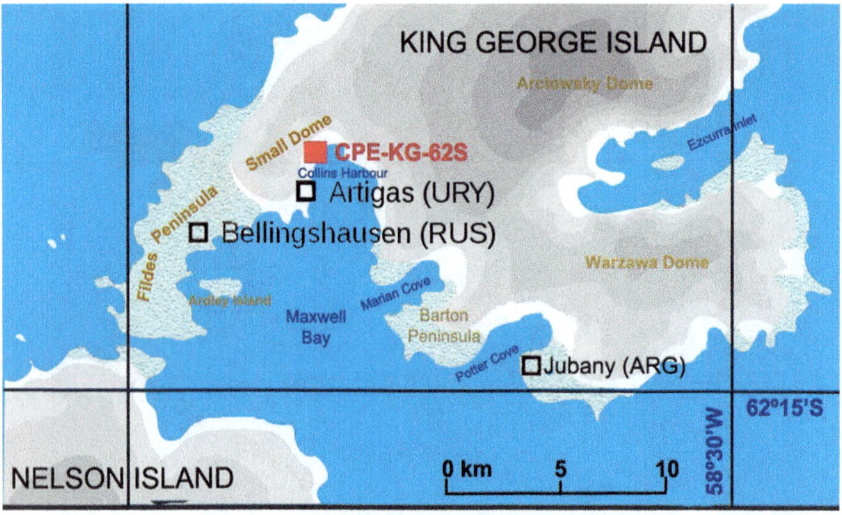

Fig. 5.2 Location of monitoring station CPE-KG-62S, next to the Uruguayan Artigas Base (*Source* Institut für Physische Geogrphie, Universität Freiburg, Germany. Laboratório de Pesquisas Antárticas e Glaciológicas, Universidade Federal do Rio Grande do Sul, Brazil)

sensitive to climate change. The ice loss in this area between 2000 and 2008 amounted to the 1.6 % of the total territory, and the glacier retreat rates are greater than observed in previous years [1].

5.2.1 King George Island

King George Island (KGI) is the largest island of archipelago (the major axis is about 75 km). KGI, between 61°54′–62°16′S and 57°35′–59°02′W, is bounded on the north by sea of Drake (place where the waters of the Atlantic and Pacific oceans are mixed); on south, the Bransfield Strait; and by the west the Bellingshausen Sea.

Eight permanent research stations and various seasonal huts and research camps are located on King George Island. Among them, Bellingshausen Russian station (62°11′ 47″S 58°57′ 39″W) and Uruguayan Artigas Base (62°11′ 04″S 58°54′ 09″W, 16 m a. s.l.) (Fig. 5.2). Bellingshausen station, named from Admiral Thaddeus Bellingshausen who explored the area in 1821, is opened since February 22, 1968, and it is a base of important aerometeorological, oceanological, glaciological, geophysical, biological and medical observations.

More than 90 % of King George Island, about 80 km wide and 30 km long (about 1,313 km^2) is covered with Collins glacier (Fig. 5.3). Ice free areas in the form of *nunataks* or inshore beaches have been formed as a result of morphological and dynamic conditions. The largest free areas occur on the southern part of the island,

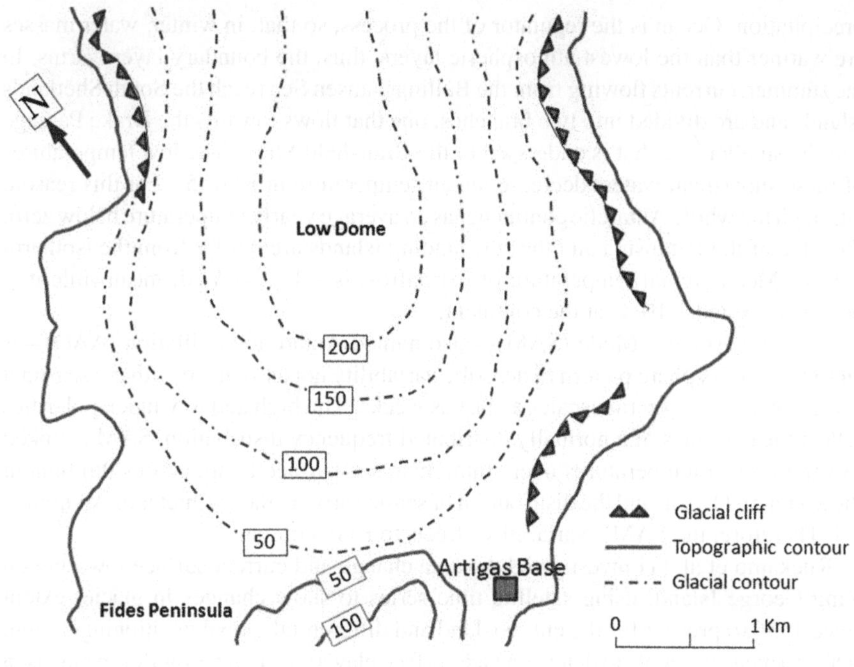

Fig. 5.3 Topographic sketch of ice thickness at Low dome also known as Bellinghausen dome, Collins glacier

where the climatic conditions are more advantageous than on the north-western coast, which is covered completely by ice. The largest *nunataks* are Fildes Peninsula, Keller Peninsula, Point Thomas region, Three Brother's Hills, and others, where the system of elevations protects the dry areas from glaciers flowing.

The coastal line in the southern part of the island, present bays formed as a consequence of tectonic predisposition and glacier activity (inland ice) [2], viz, Venus Bay, Destruction Bay, King George Bay, Admiralty Bay, and Maxwell Bay.

5.2.2 KGI Climate

The Antarctic Peninsula is one of the most rapidly warming locations on Earth, with an increasing 0.56 °C per decade measured at the Faraday/Vernadsky station over the last 50 years [3]. The Antarctic and Subantarctic regions have a prominent climatic interannual variability, compared to other latitudes, due to a number of feedbacks that result from complex interactions between atmospheric circulation, oceans and cryosphere [4].

KGI climate is determined by the passage of successive cyclonic systems, transporting relatively warm and humid air, and bringing strong winds and heavy precipitation. Ocean is the regulator of the process, so that, in winter, water masses are warmer than the lowest atmospheric layers, thus, the boundary layer warms. In the summer, currents flowing from the Bellingshausen Sea reach the South Shetlands islands and are divided into two branches, one that flows towards the Drake Passage and the smaller one that is addressed to the Bransfield Strait. The low temperatures of these superficial waters decrease the air temperature in KGI [5]. For this reason, although the whole Antarctic continent has an average yearly temperature below zero, the edge of the Peninsula and the surrounding islands are not far from the isotherm at 0 °C. Mean annual temperature of permafrost is −1 °C at KGI, meanwhile it is between −8 and −10 °C at the continent.

Southern Annular Mode (SAM)—also named Antarctic Oscillation (AAO)—is an Earth's atmosphere pattern to describe variability not associated with the seasonal cycle. SAM varies on timescales as fast as weeks. The high and low index polarities reflect the extremes of a normally distributed frequency distribution. SAM is linked to variations in temperatures over Antarctica, sea-surface temperatures throughout the Southern Ocean, and the distribution of sea-ice around the perimeter of Antarctica [6]. Therefore, the SAM's variability affects to KGI's climate.

Ruckamp et al. [1] investigated the area change and current surface lowering on King George Island, using satellite time series to show changes in glacier extent since 1956 to present for the entire island and differential global positioning system (DGPS) measurements to determine ice surface elevation changes on Bellingshausen Dome. They extrapolated the findings to the entire island in order to provide melt water budgets for individual catchments. Also, Ruckamp et al. [1] showed a strong interannual variability of two consecutive years by net accumulation measurements and forecasted the disappearance of Bellingshausen Dome by measuring surface lowering rates (1997/1998 to 2010) and the ice geometry.

South Shetlands the air temperature is characterized by a high value of variability from year to year. A distinct 5.3 years cyclicity can be found in the occurring temperatures. The mean annual air temperature meassured by Kejnak [2] over the period 1944–1996, with data from the Deception and Bellingshausen stations, was −2.8 °C. Its value varied from 0.8 °C in 1989 to −5.2 °C in 1959. The highest temperatures on the King George Island occur in the Admiralty Bay region.

5.3 Collins Glacier

Glacial geology of Collins comprises the study of the landforms and sediments created by the glaciers, both past and present.

The study area covers the south side of the ice cap Collins, commonly called dome Bellinsghausen, located in the southwest of King George Island, and near the Uruguayan Artigas Antarctic Base. Collins glacier, with 1,313 km^2 of extension, occupies almost all of King George Island, except by the south-western end, where the Fildes Peninsula is located. Collins glacier was mapped by Australian National Antarctic Research Expeditions (ANARE) from air photos taken in 1956 and 1960.

Collins glacier is named by Antarctic Names Committee of Australia (ANCA) for N.J. Collins, senior diesel mechanic at Mawson Station in 1960.

All analysis is carried out on the registered data obtained from at the station of the Insular Antarctica installed by Eraso and Domínguez [7]. The focus of this work is on Collins glacier data related to discharge and air temperature, although other time series have been also studied, as humidity, pressure, radiation, and precipitation. The data were obtained from 2001 to 2011, at monitoring station of the Insular Antarctica CPE-KG-62 °S by Eraso and Domínguez [8].

5.3.1 *Hydrological Overview*

The glacier has several domes from which the ice flows to the sea. One of the smallest is Bellingshausen Dome, located in the west end of the island, has several limbs that melt before reaching the sea, and generates nine streams that flow into both sides of the coast, by the slope of the Drake Straits in North side of the island, and through the slope of Bransfield by South. Discharge into the Bransfield is composed by five streams. After following diverse proglacier routes, the whole flow converges at a pond, where an ionospheric observation station is located. The pond output is a unique stream that flows under the bridge leading to the Uruguayan Base Artigas, and finally it reaches the sea at Maxwell Bay. Since this river is the only responsible of the glacier discharge, under the bridge (62 °11′ 03″S, 58 °54′ 41″W), GLACKMA project team installed an experimental station for monitoring the glacier, named as CPE-KG62S.

The surface of the selected glacier catchment area, corresponding to the SW sector of the Small Dome or Bellingshausen Dome, was established on the studies carried out with radio-echo sounder. However, its cartography was completed with GPS readings of the perimeter moraine and the surfacing nunataks in this SW sector of Low or Small Dome (Fig. 5.3).

The surface of the glacier catchment area (Figs. 5.1 and 5.3) is about 1.31 km^2, the development of the peripheral moraine occupies 0.25 km^2, and the fluvial catchment extends throughout 1.36 km^2. Then, the total surface of the glacier is 2.92 km^2 [9]. The maximum of the ice thickness (over 300 m) occurs in the western and central part of the King George Island ice cap (Pudelko 2003).

5.3.2 *Instruments and Facilities*

5.3.2.1 Discharge Volume Measurement

The discharge volume measurement from the glacier-fed rivers is very difficult to perform accurately due to the rough conditions and the continuous changes of flow regime. An alternative method is the indirect gauge by a pressure-measuring sensor which supplies level values. The measuring station CPE-KG62S has been equipped with register instruments of SEBA Hydrometrie Company, from Germany.

The material used at in the first years to produce the time series, consisted of a multi-parameter pressure-measuring sounder (MPS) with sensors for water temperature, conductivity and level. After continuous measuring with hourly records during two complete years, flaws in the external data-logger appeared due to the hard meteorological conditions during the winter months. This originated a non-valid register during the summer of 2003/2004, so the sounder had to be replaced by an MDS type sounder that can only measure the river level, but it is an exceptional sounder regarding its resistance in harsh and extreme conditions. MDS-Insider-2 provided by SEBA is an advanced sensor for registration of water level through a high accuracy pressure transducer, with a precision of 1 mm, in the range from 1 cm to 10 m. MSD sensor has been extended to become a multiparameter probe equipped with:

- A water temperature sensor, with a range of from −20 °C to + 70 °C and a precision of 0.01°C.
- An electrical conductivity sensor, with a range from 0 to 1 mS/cm and an accuracy value of 0.5 μ S/cm.

Sensors are protected by a cylindrical steel grating attached to probe body, a camera with a watertightness up to 150 m deep. Other ancillary material complete the station, as a data-logger to save the time series from probes with 64 kbytes memory, lithium and alkaline batteries, M-BUS Interface for laptop PC connection to download data and reprogramming, cable shielded waterproof outer, and SEBA special software for the first inspection of the raw time series generated.

5.3.2.2 Weather Measurements

About weather measurements, Antarctic environment means a hard challenge for data collection. So much so, King and Turner [4] suggest that important observations should be complemented by the ability of the human observer. So, the best option is to take meteorological information from the ground-based station of Bellingshausen (24 m asl). Data can be valid for the Collins study, due to the proximity (about 4 km). The average temperature around the station in August is −6.8 °C and +1.1 °C in February (Fig. 5.4). From the multiples variables measured in Bellingshausen base, only the following meteorological (input) parameters have been retrieved for the Collins study, though the air temperature will be the prominent parameter to consider in this thesis:

- Precipitation (P) in mm.
- Atmospheric pressure (P_a) in mbar.
- Global solar radiation (R_s) in W/m^2.
- Relative air humidity (H_r) in %.
- Air temperature (T_a) in °C. Air temperature is assumed as the main weather parameter in the Collins glacier study.

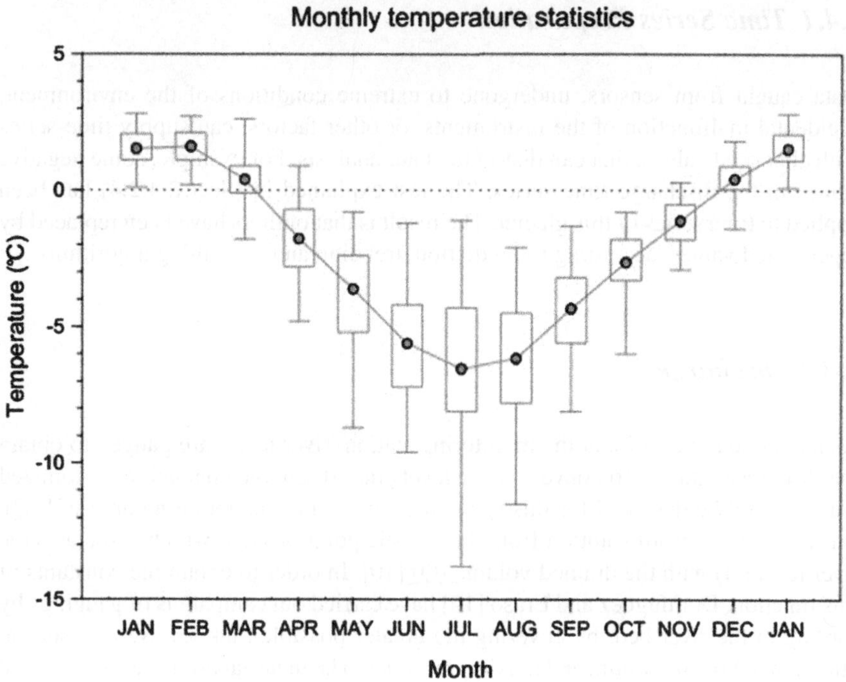

Fig. 5.4 Bellingshausen monthly temperature statistics: this plot shows the mean, quartiles and range for each month

5.4 Analysis of Time Series

In order to find causality between the weather phenomena and the glacier system, some meteorological variables have initially been considered: precipitation, air temperature, atmospheric pressure and humidity. Each one might present a different contribution to the glacier discharge. After performing XWS (Eq. 2.45) between Collins discharge and each one of the mentioned weather variables, weak correlations regarding discharge were observed between some of them; nevertheless, these extended analysis is left out of this study. The highest correlation was found between the discharge and the air temperature, although closely followed by radiation. The issue about which of the two phenomena have a more direct influence on the melting has been discussed in Sect. 3.3.1. Then, in the next sections it is assumed that the influence of the air temperature on glacier internal ablation is the direct responsible of the glacier discharge.

5.4.1 Time Series Preparation

Data caught from sensors, undergone to extreme conditions of the environment, accidental malfunction of the instruments, or other factors, can supply time series with abnormal values, that can distort the later analysis. For example, some negative values in the discharge time series. The test explained in (Sect. 2.4.2.4) has been applied to time series of this glacier. The result is that outliers have been replaced by regenerated values, according to prediction, trending and smoothing algorithms.

5.4.2 Discharge

As mentioned in (5.3.2), at the monitoring station, river levels are gauged to obtain the discharge values. To convert water levels into discharge channel, a customized curve should be designed for this specific glacier, called calibration curve f(H, Q). It is a straightforward solution from the logistic point of view, which associates the river level (H) with the drained volume (Q) [10]. In order to obtain the constants of this function, Domínguez and Eraso [10] have carried out campaigns of gaugings by fording in the river bed, by covering the greater possible rank levels, with special attention to both minimum and maximum values. The material used was an universal propeller F1 of SEBA company, which is specifically designed to this kind of tasks.

Since rivers originated from glaciers presents important variations in flow and load of sediments, before taking each measurement, the river depth should be verified. The suitable place should be a straight stretch with small variations in the cross-sectional profile of its bottom and absence of big obstacles that may distort the speed distribution of the flow. The river width is divided into sections where water speed is measured, which multiplied by section turns out the volume per time unit. The total volume will be the sum of the volumes of each section. The distribution of the flow speed in a free lamina river decreases exponentially according to the depth, from the surface to the bottom. Domínguez and Eraso [10] have established the relation between water level and discharge specifically established for the Collins glacier, with the Eq. 5.1. The correlation coefficient $R^2 = 0.99$ after painstaking works.

$$Q_g = 0.0159e^{13.98H} \tag{5.1}$$

The discharge measurements carried out in some subpolar stations are often taken with a reading interval of a day, for saving battery consumption, but the loss of information could be too important. For this reason, the values in this Collins multifunction station has been registered in time series with intervals of 1 h.

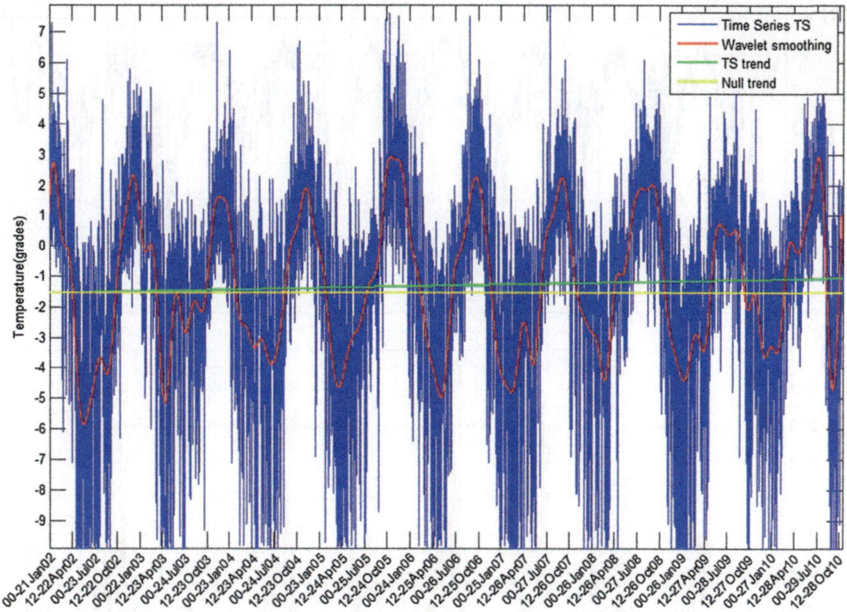

Fig. 5.5 Trend of air temperature in the period considered, with a rate of one degree per decade

5.4.3 Air Temperature

Because the ice cap is a strong heat sink, vertical and horizontal temperature gradients are constantly varying. For that reason, near-surface temperatures in the glacier are very sensitive to changes in the low layer of the atmosphere.

Figure 5.5 is the representation of the air temperature and their trend. Only in the tenth year, variations in the atmospheric temperatures reaches a half degree. The rate is about one degree per each 20 years. Blue line is the air temperature time series, the red one is the smoothed curve after applying a wavelet filter, and the green one is the linear trending. This is in agreement with King and Harangozo [11] who estimated that at the last 50 years of twenty century, the warming trend in the west sector of the Antarctic Peninsula, especially between latitudes 65 ° and 70 °S, was approximately 2 °C. The same result was obtained by Ferron et al. [12] in King George Island in the period from 1947 to 1992. Analogously, the trend of discharge has been calculated (Appendix A.5), and the result shows an increasing similar to the air temperature.

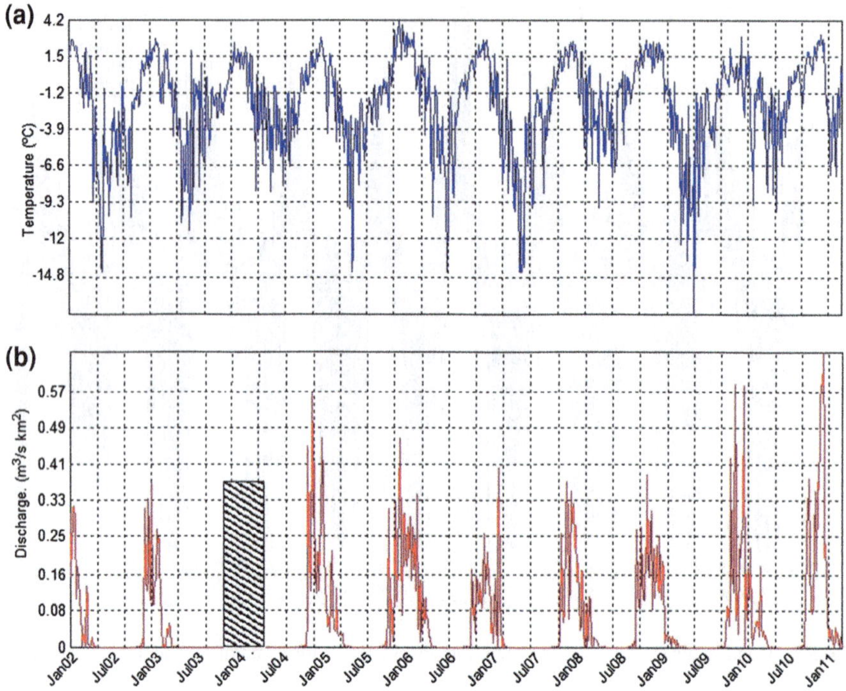

Fig. 5.6 Measurements of the runoff from glacier Collins at station CPE-KG-62S. **a** Time series of the air temperature. **b** Idem of the discharge. The third cycle data is lost for failure of sensor

5.4.4 The Involvement of Air Temperature in the Discharge

The time series of discharge and air temperature range from January 21, 2002, until May 31, 2011, with one-hour sampling periods (Fig. 5.6). Both time series show a strong yearly cyclic behavior. In cold periods, obviously there are not observable effects in the discharge time series, since there is no flow due to a complete freeze. Therefore, the dynamics is going to be studied only in periods when the glacier is active, i.e. in those which the discharge presents a significant reaction to the air temperature. In catching data in general, due to hard conditions of environment or other factors, sensors can provide raw data with some questionable values, i.e. outliers (Sect. 2.2.2). The two series, that are going to be treated in the following sections, require a prior preparing work to check the robustness of the data and perform the proper correctness. The criterion to inquire and revise outliers in air temperature and discharge time series has been the Wavelet-Rosner test (Sect. 2.4.2.4), that also was applied in karst aquifer discharge by Chinarro et al. [13].

The Fig. 5.6 displays the cycle 3 corresponding to the 2013–2004 season hidden by a shaded shape, given that data were lost that year for a sensor breakdown.

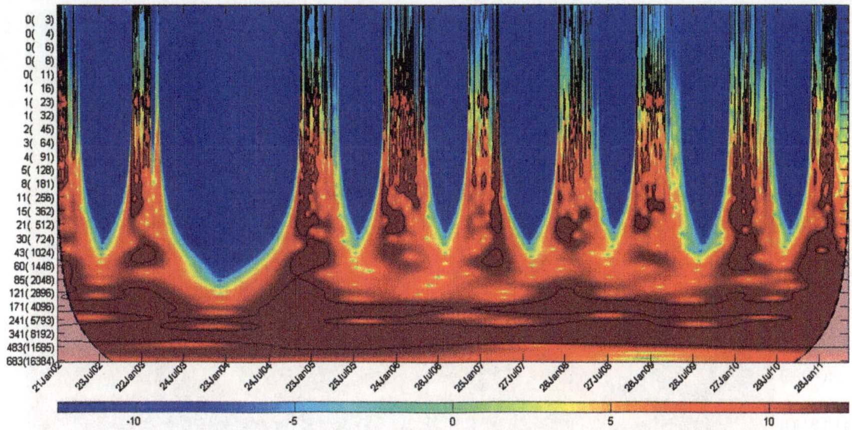

Fig. 5.7 Discharge wavelet spectrum at the Collins glacier. The *vertical axis* has a logarithmic scale in days (hours). The *abscissa axis* represents the time line labeled quarterly. Values of the spectral power are depicted by a colour map. The cone of influence is also represented to indicate the useful area of the spectrum without influence of the edge effects. The 2003–2004 winter is empty due to loss of data

5.4.5 Time Series Power Spectrum

The graphical output of Fig. 5.7 has been generated by a computer application specifically designed for this analysis, with algorithms based on the work of Grinsted et al. [14]. The basis function for the wavelet transform is a Morlet mother, commonly used in hydrology [13] because it permits a reasonable precision in situating at timeline, the signal frecuencies [15, 16]. Lafrenière et al. [17] also uses the Morlet wavelet for analysis of runoff regimes of a glacial watershed.

In the ordenate axis, numbers outside the parentheses indicate the equivalent in days. The abscissa axis represents the timeline labeled quarterly, although the calculation of spectral power is held every hour. Values of the spectral power are represented by a distribution of color hue depending on the moment in time series and the frequency (or scale). The cone of influence indicates the useful area of the spectrum without influence of the edge effect.

Both spectrograms, discharge (Fig. 5.7) and temperature (Fig. 5.8), show the annual cycles, identified as a horizontal band of high power over 365 days. The discharge also has another high power band corresponding to six months. This band is significantly less powerful in the temperature spectrogram. This is related with the fact that the glacier changes its activity every six months: completely frozen to active and vice-versa. All these results are corroborate with Braithwaite and Olesen [18]. Both spectrograms also present a richer spectrum with frequency components in the band of about 51–60 days. There is another high power band in the range of 21–43 days. The reason for these bands is not well understood, and in a further research based on these facts could throw new findings.

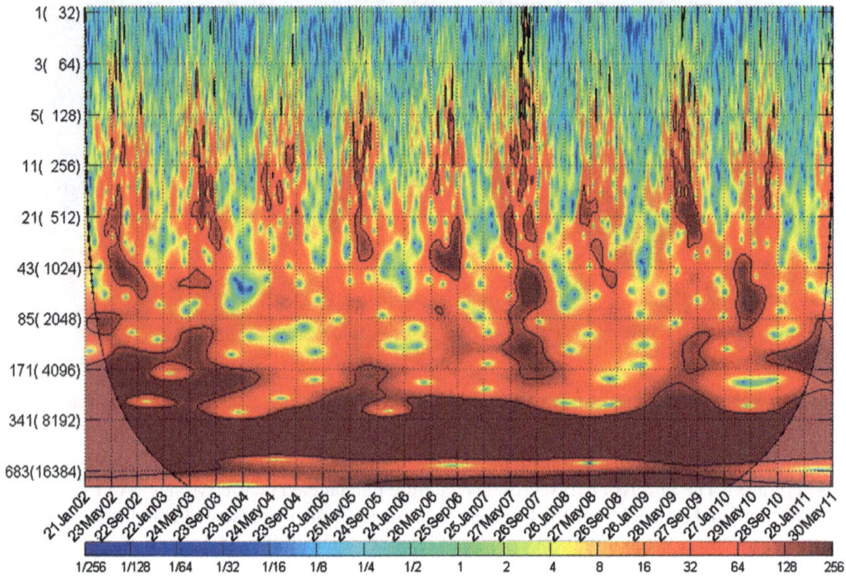

Fig. 5.8 Air temperature wavelet spectrum at the Collins glacier

5.4.6 Periods of the Annual Cycle

The structure of discharge time series observed comprises high discharge periods of the glacier that are alternated with winter periods in which low temperature keeps the solid state of the glacier and yields null values in the sensor measurements. The differences observed between the correlograms of the temperature (system input) and the discharge (system output), are also highlighted in the wavelet coherence spectrum. In cold periods, the coherence has no high frequency components because the glacier is completely frozen, although temperature presents diverse frequency components in its power spectrum. Nevertheless, when the glacier is active—in summer—the coherence is very high.

Since the cycling behavior has clearly been manifested in the coherence representation, The next stage is to show that a model based on an annual cycle can represent, up to a certain point, any period in the time range considered. Assuming statistically that the probability distribution of a cycle is preserved along different years, a disaggregation study can be posed according with a significant body of hydrological literature [19]. Thus, three periods can be observed in one annual cycle (Fig. 5.9), according to the response of the discharge to the air temperature:

(a) *Inactive Period* occurs during the winter time when very low temperatures prevent the discharge, and the system dynamics remains detained. In this period, temperature and discharge are in-coherent.

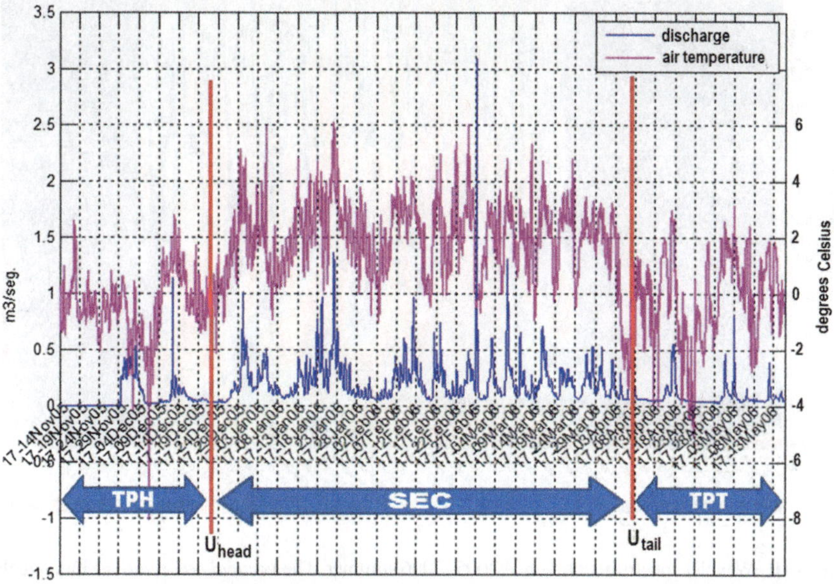

Fig. 5.9 Approach of inactive, effective and transition periods boundaries of discharge and air temperature signals for one annual cycle

(b) *Effective Period*, that corresponds with the core of the summer when all the conduits in the glacier are well formed. It bounded by separators U_{head} and U_{tail} to form the Seasonal Effective Core (SEC) of the annual hydrological cycle. SEC can be noted by the existence of water flow at the data collection point. Nevertheless, in the estimation of SEC, a better precision is required to assure the quality of the system model. Now the posed problem is to find an analitic method for detecting and quantifying, in each seasonal cycle, thresholds (U_{head} and U_{tail}) delimiting the core, inside which signals are supposed to exhibit a certain dependency, as is shown below. In this period temperature and discharge present high coherence.

(c) *Transition Period* includes all phenomena that occur between warm and frozen seasons in each annual cycle. At the end of the winter, solar radiation gradually begins to be stronger and consequently, the air temperature increases. The time series of discharge may be noticed on this stage as an irregular response to the air temperature. Moreover, the response can become so complex that some abrupt variations can be observed in the time series of the output drainage, without apparent correspondence with a peak in the air temperature signal (zone preceding the separator U_{head}. So, in the transition period, the coherence between temperature and discharge is weak and unstable. There are two transition periods, *Transition Period* before the summer, and *Transition Period Head* (TPH) after summer. The transition periods are characterized by snowmelt processes

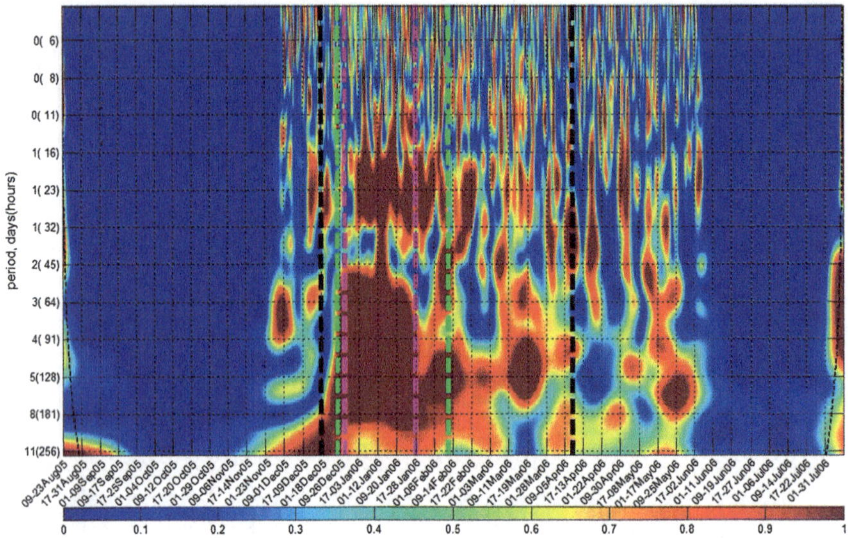

Fig. 5.10 Wavelet spectrum coherence 2005–2006 summer. The *vertical colored lines* indicate the limits of the head and tail of the SEC

produced by the absorbed energy, and comprises three phases: *warm-up*, *reservoir* and *release* according with [20, 21].

The main interest of the present work is in estimating the Effective Period, called hereafter *Seasonal Effective Core* (SEC), i.e. the time when glacier discharge responds in a coherent way to the temperature. The SEC of each cycle is the basis to build mathematical models of the glacier. The next section presents the analytical processes to determine the SEC boundaries.

5.4.7 Seasonal Effective Core Calculus

In this section, we propose a method of calculus of the separator positions based on the coherence exhibited by temperature and discharge signals.

The wavelet coherence spectrum (Appendix A, Fig. A.6) between temperature and discharge time series confirms the cyclical distribution (yearly) of the coherence along the whole period under study.

The coherence spectrum for a seasonal cycle is depicted in the Fig. 5.10 during the 2005–2006 summer. The SEC can be distinguished by a high coherence in the central area of the spectrum, manifested by an increased density of dark red clumps. The vertical colored lines indicate the limits of the U_{head} and U_{tail} of the SEC depending on the coherence level. The black one delimits an area with more than 0.70 coherence. The green lines are the bounds for an average coherence more than

0.75. Inside the area delimited by magenta lines, the average coherence is more than 0.80 and this area is considered the core of the effective period. The next issue is to estimate the limit values or thresholding than define the vertical line equation and bounds the SEC. A pitfall is that the coherence thresholds can not be rigorously represented by a vertical straight line because the WPC is composed of different frequency bands. Following the time line, a given value of coherence can be reached in one band before the others. With the coherence equation (Sect. 2.7.2.3), the scale and a threshold h_j (for $h_j = 0.5, 0.55, 0.6, 0.65, 0.7, 0.75, 0.8$), the set of points describing the bounds of coherence can be obtained for each scale s_i. A recursive algorithm has been developed to estimate formal bounds.

The frequency spectral distribution is simplified in an univariate function that determines clearly the cutoff points, and, consequently, the coherence thresholds. The result is the Coherence Average Function (CAF) obtained by mapping the tridimensional distribution of coherence in a bidimensional function. i.e., CAF reduces the WPC, which depends on time and frequency, to a mono-variable function that depends only on time by calculating a statistic estimator ($\mathbb{E}$), expressed in:

$$\Gamma_{\psi}[u(t), y(t)]_{s,\tau} = \mathbb{E}(\mathscr{C}_{\psi,s,\tau}\{u(t), y(t)\}) \tag{5.2}$$

The statistic estimator chosen is the root mean square (RMS), due to easy implementation and slight differences with other estimators tested. The expression (5.3) calculates the coherence of each spectral band at a given time.

$$\Gamma_{\psi}[u(t), y(t)]_{s,\tau} = \sqrt{\tfrac{1}{S_2-S_1}\int_{S_1}^{S_2}\left(\mathscr{C}_{\psi,s,\tau}\{u(t), y(t)\}\right)^2 ds} \tag{5.3}$$

However, the CAF is subject to substantial fluctuations and does not establish a well-defined border point for the SEC, as is shown by the red curve in the last plot of Fig. 5.11. The levels of coherence are set by the horizontal coloured lines, with values from 0.5 to 0.8. To solve this problem, a moving average is applied to the CAF, obtaining the *Smoothed Coherence Average Function* (SCAF). The SCAF is plotted in the magenta curve of Fig. 5.11. This gives a better definition of the crossing points, avoiding ambiguities due to the excessive ripple in the CAF signal. When the SCAF crosses the desired coherence level, U_{head} and U_{tail} are established. In the present work, the 0.8 level of coherence is used to establish the SEC of all the cycles. With this level of coherence, for all the cycles studied, the SECs do not exhibit sudden water releases or flow detentions not related with the temperature changes. Table 5.1 shows the different SEC,s for each cycle at this coherence level.

To illustrate the process of the estimation of each yearly SEC, cycle 2005–2006 in the first plot of Fig. 5.11 shows the air temperature and discharge of this cycle, when the glacier is active. TPH, SEC and TPT can be observed bounded by separators U_{head} and U_{tail}. The WPC for this cycle is depicted in the second plot of Fig. 5.11. The SEC can be distinguished by a high coherence in the central area of the WPC, manifested by an increased density of dark red clumps. The vertical limits U_{head} and U_{tail} of the SEC depend on the coherence level. The upper plot in Fig. 5.11 shows

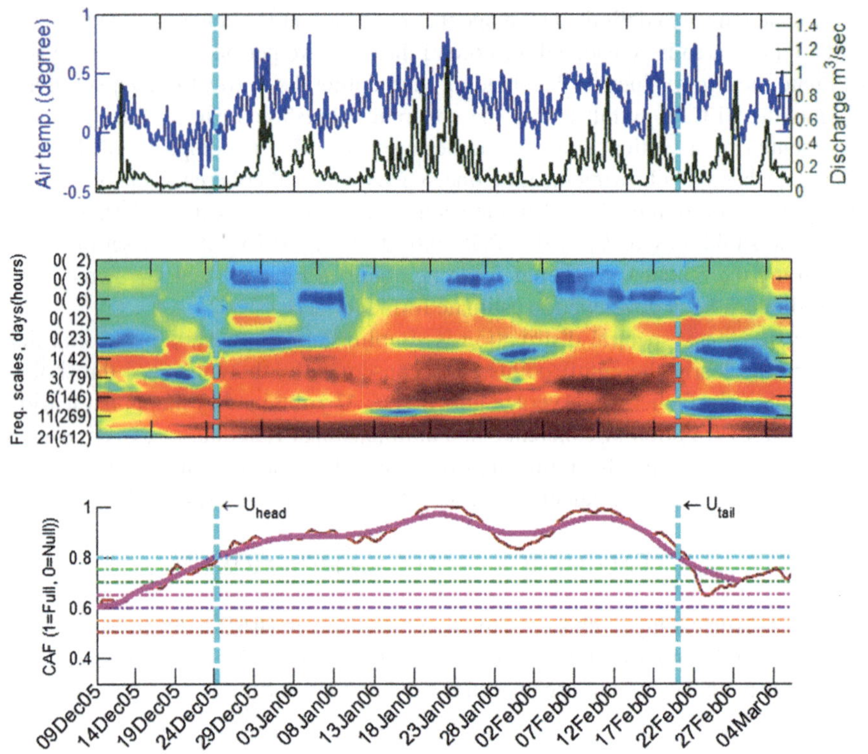

Fig. 5.11 Time series, wavelet coherence spectrum, and coherence levels in the 2005–2006 cycle of Collins glacier

the time series; the middle one the WPC and the bottom one the CAF and SCAF. Horizontal coloured lines indicate levels of coherence from 0.5 to 0.8. The red curve is the CAF and the magenta curve is the SCAF. The crossing points of the horizontal coherence line levels with the SCAF curve determine the vertical lines of colour, indicating the limits of the head and tail of each SEC.

5.5 System Identification

One of the main objectives of the present chapter is to obtain a mathematical model of the hydrological dynamics of the glacier, after assuming that the temperature is the principal cause of the discharge variability. However, this is only certainly feasible in the SEC of each hydrological cycle. Thus, the work is focused on obtaining a model of the glacier activity during the SEC, having excluded the study in transition periods.

Table 5.1 Limits of the SEC for all cycles for a coherence level of 0.8

Cycle	SEC limits
2001–2002	*Jan26–Feb09*
2002–2003	*Dec26–Feb19*
2003–2004	(n/a)
2004–2005	*Dec23–Feb15*
2005–2006	*Jan09–Feb13*
2006–2007	*Dec30–Feb14*
2007–2008	*Feb01–Mar06*
2008–2009	*Dec01–Mar25*
2009–2010	*Jan11–Mar08*
2010–2011	*Dec14–Feb05*

Because the physical processes that govern the glacier behaviour are too complex to permit a sufficiently accurate modelling level, the glacier is assumed as a *black box* system. This assumption has facilitated the parametric identification, and the oversimplification, in taking a unique time series for the input (temperature) and another for the output (discharge), has been enough to reveal interesting correlated properties. To bear out the identification of the glacier, two different hypotheses have been addressed.

1. To consider the glacier as a linear system. Linear parametric identification techniques were used to obtain a transfer function of the glacier system.
2. To consider that some nonlinearities remain in the glacier SEC. The Hammerstein-Wiener technique was applied to isolate and to identify the nonlinearities.

Firstly, the models obtained with the two approaches were compared to establish their relative accuracy. Secondly, in order to analyze how the glacier dynamics is evolving, the SEC boundaries were estimated per each cycle to obtain an individual annual model. Then, a model, obtained in this way, was compared to each others. Finally, based on blending of all SECs, a global model of the glacier was obtained: the Coalescent Model.

5.5.1 Linear Model by Cycles

In black-box parametric identification, a set of generic standard structures is taken into account [22]: Output Error (OE), Finite Impulse Response (FIR), AutoRegressive with eXogenous terms (ARX), AutoRegressive-Moving Average with eXogenous terms (ARMAX) and Box-Jenkins (BJ). Also, for each structure, different order transfer functions are tested. The couple function-structure that obtains the best efficiency, according to Eq. 2.51, was chosen. This systematic work has been carried out for the identification of the Collins glacier. In fact, the differences in quality between the tested models are not relevant. For this reason, the most simple model, the Output

Table 5.2 OE linear model efficiency. M1, …, M9 refer to the models created on known data in each year

M_1	M_2	M_3	M_4	M_5	M_6	M_7	M_8	M_9	M_{10}
0.61	*0.37*	–	*0.72*	*0.63*	*0.61*	*0.78*	*0.59*	*0.39*	*0.62*

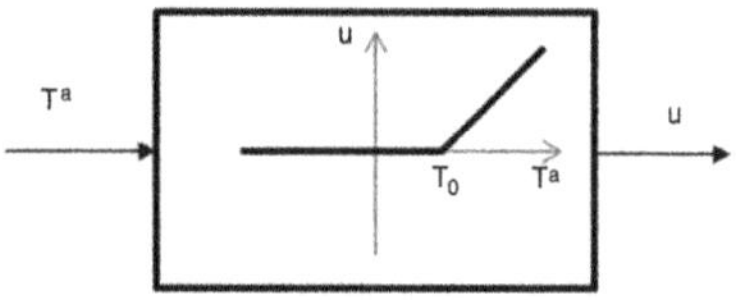

Fig. 5.12 The Hammerstein block is a dead zone with a breakpoint in the temperature T_0

Error (OE) with two poles and one zero, has been used. This simplification makes the subsequent explanation and further analysis easier with a minimal loss of accuracy.

Table 5.2 shows the quality of the models obtained with each SEC (M_j) expressed in terms of efficiency (Eq. 2.51). As can be seen, all the models present quite poor values. Only M_4 and M_7 have E greater than 0.65. So, the hypothesis of linearity seems not to be successful, although only the SEC has been used to identify each cycle. This requires new approaches considering that the system is not linear at all.

5.5.2 Nonlinear Model by Cycles

To try to improve the quality of the resulting models, the nonlinear Hammerstein-Wiener model structure has been used, maintaining OE as the linear block. A large set of static nonlinearities has been tested for the input and output blocks. The conclusion of these tests is that the best structure is a dead zone for the input and an identity for the output. The dead zone is in the range $(-\infty, T_0]$ as can be seen in Fig. 5.12. The T_0 that gives the maximum efficiency for each cycle ranges from 0.7 to 2.0 °C, with a mean close to 1.5 °C. For this reason, in order to be able to compare all the models, T_0 has been fixed at 1.5 °C. This kind of nonlinearity seems to be related to the state change from ice to liquid water. The gap between zero degrees and T_0 seems to be due to the fact that the temperature is measured at the Bellingshausen base, located on bare ground about two kilometres away from the glacier and in a place with low albedo, with an overheating of the area.

Table 5.3 shows the quality of the models (efficiency according to Eq. 2.51) of each cycle obtained with the previously described dead zone. As can be seen, the quality of all models has improved over the corresponding linear models. Except for M_9, all model efficiencies are over 0.65. Model based on cycle nine is a special case since the very low coherence level of the SEC, probably due to a particularly snowy summer. Therefore, the model based on nineth cycle is mathematically unstable, and they are not going to regard in the study henceforth.

Table 5.3 Hammerstein-Wiener model efficiency. M1, ..., M10 are models based on recorded data of years 2001,...,2011, respectively

M_1	M_2	M_3	M_4	M_5	M_6	M_7	M_8	M_9	M_{10}
0.67	*0.75*	–	*0.81*	*0.73*	*0.68*	*0.80*	*0.71*	*0.51*	*0.77*

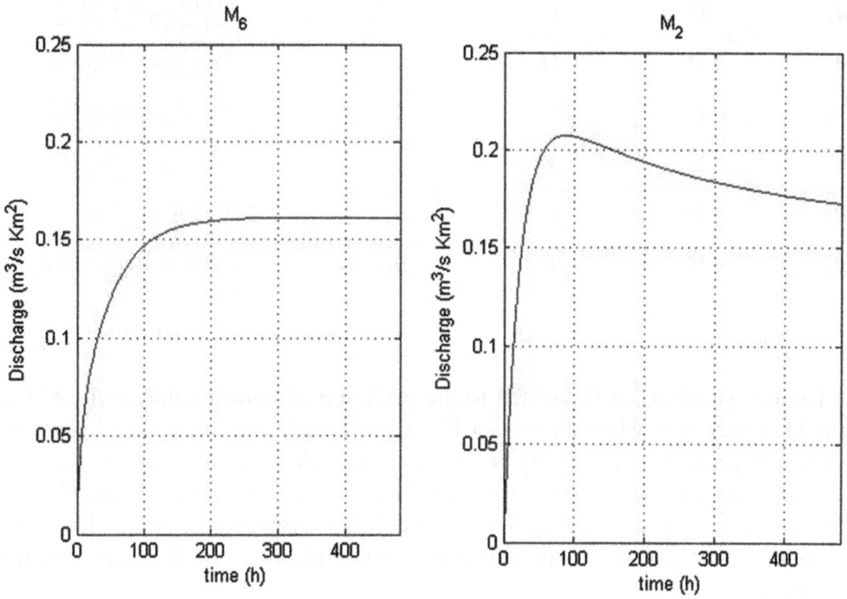

Fig. 5.13 Two different step responses. M_6 has a typical overdamped response (M_1, M_5 and M_8 have similar efficiency). However, M_2 has a non negligible overshoot (M_4, M_7 and M_{10} have similar efficiency)

Table 5.4 Average temperature in Celsius degrees

	1	2	4	5	6	7	8	9	10
Annual	0.88	−2.85	−1.59	−1.55	−2.0	−3.04	−1.04	−3.13	−1.58
SEC	2.46	2.05	1.33	2.65	2.09	1.91	1.7	1.12	1.2

The identified models present two main types of responses, as can be seen in Fig. 5.13. On one hand, models M_1, M_5, M_6 and M_8 have a step response that corresponds to a basic overdamped second order system. On other hand, models M_2, M_4, M_7 and M_{10} have an overshoot caused by the zero present in the second order model. Among some hypotheses analyzed to explain such behavior, it is worth mentioning the one related to the annual cycle and the SEC average temperature. Models that show overshoot correspond to very cold winters or cold summers. Table 5.4 shows the measured average temperature of annual cycles and SECs.

Table 5.5 Characteristic parameters of the step response year by year and linear transfer function

Model	Static gain	Rise time	Overshoot	Transfer function Y(z)/X(z)
M_1	0.25	8	0	$\frac{0.0625z^{-1}-0.02569z^{-2}}{1-1.152z^{-1}+0.2993z^{-2}}$
M_2	0.16	23	0.31	$\frac{0.009932z^{-1}-0.009904z^{-2}}{1-1.952z^{-1}+0.9523z^{-2}}$
M_4	0.17	1	2.01	$\frac{0.01973z^{-1}-0.01973z^{-2}}{1-1.949z^{-1}+0.9494z^{-2}}$
M_5	0.15	11	0	$\frac{0.0265z^{-1}+0.01514z^{-2}}{1-0.4069z^{-1}-0.3241z^{-2}}$
M_6	0.12	69	0	$\frac{0.01003z^{-1}-0.009464z^{-2}}{1-1.808z^{-1}+0.812z^{-2}}$
M_7	0.27	39	0.83	$\frac{0.008113z^{-1}-0.008103z^{-2}}{1-1.985z^{-1}+0.9854z^{-2}}$
M_8	0.29	170	0	$\frac{0.02914z^{-1}-0.02845z^{-2}}{1-1.761z^{-1}+0.7631z^{-2}}$
M_{10}	0.19	40	0.37	$\frac{0.008883z^{-1}-0.008889z^{-2}}{1-1.979z^{-1}+0.9787z^{-2}}$

The overshoot could be related to the evolution of conduits during the SECs. In cold summers or when the glacier has a low temperature after a cold year, the conduits develop slowly and may be partially closed due to re-freezing. An increase in temperature provokes the re-opening of the conduits and fast emptying flow. In order to analyze if the dynamics of the glacier is evolving, Table 5.5 shows the characteristic parameters of the unit step response of all the models. The selected parameters are:

1. *Static gain*. Ratio of the system output and the system input values under the steady state condition.
2. *Rise time*. The time required, in hours, for the output to reach 90 % of the steady state final value.
3. *Overshoot*. Ratio of the maximum peak value (minus the steady state final value) and the steady state final value.

In other hand, no detectable trend can be seen in the data, and the different types of years are interleaved. It seems that the series is too short to detect trends.

5.5.3 Glacier Coalescent Model (GCM)

The GCM of the Collins glacier tries to represent a general behaviour of the glacier system by blending all *seasonal effective cores*, as given by Eq. 5.4:

$$\Upsilon = \bigcup_i (SEC)_i \tag{5.4}$$

Table 5.6 Eficiencies of overdamped and overshoot models

Model	Auto-simulation	Prediction C_8	Prediction C_{10}
Overdamped	0.63	0.42	0.21
Overshoot	0.48	0.41	0.55

The SEC of cycles C_1, C_2, C_4, C_5, C_6, C_7 and C_8 are used to identify the GCM. The SEC of cycle C_{10} is used to validate the prediction power of the model. As in the previous section, the structure for identifying the GCM is a Hammerstein-Wiener model with OE for the linear block and a dead zone in the input, with $T_0 = 1.5\,°\mathrm{C}$. The GCM obtained presents an efficiency E = 0.40 in the simulation of cycles used for the identification and an efficiency E = 0.41 in the prediction of the discharge of cycle C_{10}. Both are indeed poor efficiency.

Following the analysis performed on the step response of the cycle models, the cycles are classified as overdamped or overshoot. For this reason, it seems reasonable to divide the data into two different sets, those belonging to the overdamped cycles and those belonging to the overshoot cycles. Thus, two different coalescent models are identified: The overdamped (Eq. 5.5) and overshoot (Eq. 5.6)

$$\Upsilon_{overdamped} = \bigcup\{C_1, C_5, C_6\} \tag{5.5}$$

$$\Upsilon_{overshoot} = \bigcup\{C_2, C_4, C_7\} \tag{5.6}$$

Cycles C_8 and C_{10} are used to analyze the prediction power of the models. The results of this last identification process are presented in the Table 5.6

As suspected, the GCM of the overdamped cycles predicts the dynamics of SEC of C_8, overdamped, better than the SEC of C_{10}, overshoot. The same happens with the overshoot GCM that better predicts the overshoot cycle C_{10}. However, the efficiencies still have a low value reflecting the difficulty of finding a model that represents the general behaviour of the glacier, at least with the classical techniques used here.

5.6 Conclusion

In the present chapter, the relationship between the water discharge of the Collins glacier, located on King George Island, Antarctica, and the local air temperature has been studied from the System Engineering standpoint. Time series taken from level sensors and weather station Bellinghausen base from January 2001 to April 2011 have been the base of the study. The wavelet analysis was used to establish the correlation and coherence between temperature and discharge.

The wavelet spectrum coherence between the air temperature and the discharge shows that the two variables are not always related in the same manner. Thus, three different periods have been established: (1) the *Inactive Period* occurs during the winter time, when very low temperatures prevent the discharge at all, and the system dynamics remains detained (temperature and discharge are incoherent); (2) the *Effective Period* corresponds with the warm season when all the conduits in the glacier are well formed (temperature and discharge present high coherence); and (3) the *Transition Period* occurs between the warmer and freezing seasons in each annual cycle (the discharge has an irregular response to the air temperature because of frequent episodes of opening and closing of internal ducts). An analytical technique based on the wavelet coherence density has been proposed to determine the boundaries of each period, here called the *Seasonal Effective Core* (SEC) because it refers to the period when the glacier discharge responds in a coherent way to the temperature. The rest of the study to identificate the model of the glacier is focused on the SEC.

The linear parametric identification technique was applied to each glacier SEC to obtain linear models. However, the quality of the models obtained were not good, and thus the hypothesis of linearity was discarded. To take into account the nonlinearities of the glacier system, the non-linear Hammerstein-Wiener model structure was tested and compared to other models. On one hand, the identification work with this technique has revealed that one of the main nonlinearities present in the glacier behavior is the phase change of water. This mathematical conclusion obviously agrees with the physical reality. On the other hand, two different types of glacier dynamics have been discovered (overdamped and overshoot) depending on the annual cycle and the SEC average temperature. Models that present overshoot correspond to very cold winters or cold summers.

Parametric identification has been proved to be a powerful analysis technique, but its predictive power is insufficient. Further research is required to validate other nonlinear identification techniques, such as, for example, neural networks, in order to obtain more accurate predictive models.

References

1. Ruckamp M, Braun M, Sukro, Blindow (2011) Observed glacial changes on the King George Island ice cap, Antarctica, in the last decade. Global Planet Change 79(1–2):99–109
2. Kejnak M (1999) Air temperature on King George Island, South Shetland Islands, Antarctica. Pol Polar Res 3:183–201
3. Turner J, Colwell SR, Marshall GJ, Lachlan-Cote T, Carelton AM, Jones P, Lagun V, Reid P, Iagovkina S (2005) Antarctic climate change during the last 50 years. Int J Climatol 3:279–294
4. King J, Turner J (1997) Antarctic meteorology and climatology. Cambridge University Press, Cambridge
5. Jiahong W, Jiancheng K, Zichu X, Jiankang H (1994) Climate, mass balance and glacial changes on small dome of Collins Ice Cap, King George Island,Antarctica. Adv Polar Sci 5(1):52–61
6. Thompson DJ, Wallace JM (2000) Annular modes in the extratropical circulation. Part i: month-to-month variability. J Clim 13(5):1000–1016
7. Eraso A, Domínguez C (2001a) Subpolar glaciers network as natural sensors of global warming evolution. Ministerio de Medio Ambiente, Instituto Nacional de Meteorología, Conferencia de Clausura de la V Edición del Curso Magíster de Riesgos Climáticos e Impacto Ambiental, p 16
8. Eraso A, Domínguez C (2001b) Substantial changes happened during the last years in the icecap of King George, Insular Antarctica. Stud Fac Earth Sci, Univ Silesia 45:87–110

9. Domínguez C, Eraso A (2007) Karst and Cryokarst. Department of Geomorphology University of Silesia, Frequent systematic errors in the measurement of the glacier discharge
10. Domínguez M, Eraso A (2009) Substantial changes happened during the last years in the icecap of King George. Technical report, PANGAEA—Publishing Network for Geoscientific & Environmental Data, Insular Antarctica
11. King J, Harangozo S (1998) Climate change in the western antarctic peninsula since 1945: observations and possible causes. Annu Glaciol 27:571–575
12. Ferron FA, oes JCS, Aquino FE, Setzer AW (2004) Air temperature time series for King George Island, Antarctica. Braz Antarct Res 4:155–169
13. Chinarro D, Villarroel J, Cuchí J (2011) Wavelet analysis of Fuenmayor karst spring, San Julián de Banzo, Huesca, Spain. Environ Earth Sci, Special Issue:1–13
14. Grinsted A, Moorel JC, Jevrejeva S (2004) Application of the cross wavelet transform and wavelet coherence to geophysical time series. Nonlinear Process Geophys 11:561–566
15. Foufoula-Gergiou E, Kumar P (1994) Wavelets in geophysics, (Wavelet analysis and its applications), vol 4, 1st edn. Academic Press, Boston
16. Schaefi B, Maraun D, Holschneider M (2007) What drives high flow events in the Swiss Alps? recent developments in wavelet spectral analysis and their application to hydrology. Adv Water Resour 30:2511–2525
17. Lafrenière MM, Sharp AK (2003) Wavelet analysis of inter-annual variability in the runoff regimes of glacial and nival stream catchments, Bow Lake, Alberta. Hydological process 17:1093–1118
18. Braithwaite R, Olesen O (1988) Effects of glaciers on annual run-off. Johan Dahl Land. South Greenland. J Glaciol 34(117):200–206
19. Santos EG, Salas JD (1992) Stepwise disaggregation scheme for synthetic hydrology. J Hydraul Eng 118(5):765–784
20. Katsushima T, Kumakura T, Takeuchi Y (2009) A multiple snow layer model including a parameterization of vertical water channel process in snowpack. Cold Reg Sci Technol 59:143–151
21. Jouzel J, Souchez R A (1982) Melting-refreezing at the glacier sole and the isotopic composition of the ice. J Glaciol 28(98):35–42
22. Ljung L (1999) System identification.Theory for the user, 2nd edn. Prentice Hall, Upper Saddle River

Chapter 6
Final Conclusions: Synthesis, Contributions and Perspectives

Amicus Plato amicus Aristoteles magis amica veritas.
Isaac Newton (1642–1727)

6.1 Introduction

The expectations of any Ph.D. thesis should be that it contains something novel as its centerpiece and that it is confined to a narrow field of knowledge. In addition, it should be a documentary evidence of the author's ability to study in detail a particular approach to a problem, as well to lead and bring to fruition a large-scale research project with an international scope, wide impact and high innovative load.

This chapter shows that the expectations of this thesis have been fulfilled. After presenting the objectives, techniques, natural system features under study as well as data processing and analysis in the experimental cases, it is necessary to highlight the features of the findings to round off the dissertation.

The most effective way to conclude this thesis is to synthesize the significant thoughts that the author has developed. These final pages provides the last chance for the reader to assess this work as a serious effort in contributing to system identification and hydrology science, as well to evaluate the author's ability to organize and conduct leading and profound research work.

> The purpose of this last stage is to integrate the various issues covered throughout the previous chapters, highlighting the findings already set out, with additional comments upon the meaning of all of it, the implications resulting, answers given to the hypotheses posed at the beginning, and to raise possible working lines open for future research.

D. Chinarro, *System Engineering Applied to Fuenmayor Karst Aquifer (San Julián de Banzo, Huesca) and Collins Glacier (King George Island, Antarctica)*,
Springer Theses, DOI 10.1007/978-3-319-08858-7_6

Outline of this chapter
Section 6.2 is a **synthesis** that captures of the key points in the development of this thesis, and the main questions raised at the introductory chapter, which have been solved.

Section 6.3 is a **summary** of the core chapters, with an abridgment of the issues treated in each one.

After collecting data, performing analysis and achieving results, Sect. 6.4 presents this thesis **findings**, from personal considerations and insights, which have extended my knowledge on the system under study.

Section **contributions** provides an overview of the scientific contributions that over the four years of this thesis have been developed, from which most of the results have been published in conference proceedings or scientific journals.

Section 6.6 consists of remarks on the analysis processes and **discussion about the results** obtained and some caveats about interpretation of them.

Perspectives in Sect. 6.7 is an orientation toward future works that can stem from this thesis.

Finally, as a colophon, Sect. 6.7.2 is a **final remark** on nonlinear system identification applied in natural systems.

6.2 Synthesis

Systems science, which comprises the observation, identification, description, experimental investigation and theoretical explanation of the facts through laws of physics, has been applied here to interrelationships and correlation associated with two kinds of natural phenomena.

This thesis has addressed systems engineering abstract elements, using the methodologies of systems science, in order to conceive mathematical models for simulation and prediction, which have been used in the identification and analysis of hydrology systems.

Mathematical model techniques have proved their utility in understanding the karst aquifer and, consequently, better management, forecasting and control of groundwater systems. Also in a glacier system, a better interaction identification between climate and glacier discharge can be achieved through models.

System identification is a powerful tool to build accurate models of complex systems from noisy data, that has been used in many applications of various scientific disciplines, although under different terminology. From the engineering methods, the identification theory has provided a systematic methodology to choose the suitable mathematical model or adjust it to the system behavior under study. The features and skill of linear and nonlinear methods of system identification have been harnessed in natural systems. Model identification has been carried out in an iterative procedure, which involves choosing a model structure, parameter estimation or spectral assessment and model validation.

The overview through a variety of modeling approaches has made it possible to establish the necessary context for discussing identification methods and the suitability between the type of model and system requirement, in order to obtain a simulation with the best tuning. The applied methodology comprises both linear and nonlinear in order to initially test alternatives for generalized or simplified identification procedures, and, on the other hand, the detailed analysis through iteration algorithms, time-frequency transforms, and grouped models, required in general in nonlinear systems. Among them, the LTI function transfer and *z-transform* have been chosen for linear model processes. The presence of nonlinearities in the observed systems poses challenges that should be dealt with by employing methods stemming from block model structures and analyzed with wavelet techniques.

In order to explain many imperceptible events of natural phenomena, non-stationary series analysis are generally an attractive issue for various research areas. The wavelet transform techniques, widely used in last two decades, provide better results than former techniques for the examination of Earth science phenomena and for the feature detection of signals evolution. In this study, wavelet spectrum analysis offers a time-frequency panorama for weather variables related to discharge. In addition, nonlinear and linear methods grouped in a Hammerstein-Wiener structure were applied to two geographical locations with different climates and characteristics, with satisfactory results.

6.3 Summary of the Main Chapters

While the first chapters dealt with general introductions and techniques, Chap. 3 provided a conceptual explanation of karst and glacier systems, to set goals of analysis and fundamentals of the proposed model. Also, it highlighted the difficulties in designing physical models when there is neither information on the structure of the system nor physical laws that underlie behavior. In these cases, a black-box model is a reasonable option.

Karst is a landform, caused primarily by dissolution of limestone, with characteristic landscapes manifested in several ways as caves, deep sinkholes, sunk streams, and diverse springs. Abundant karst systems in Guara Mountains have attracted scientific interest to identify and characterize a hydrological model.

The glacier conceptual models related to the melting process and water discharge of a glacier have been examined in the environment of King George Island in Antarctica. The coupling between climate and glacier systems has been explained by the melting phenomenon that relates parameters from climate system, and discharge belonging to the hydrological glacier system. Therefore, this process is a hint of that subpolar glacier could have been a witness to weather changes in the last decade, due to their sensibility in the melting process variations.

6.3.1 Analysis and Identification of Fuenmayor Aquifer

Fuenmayor's spring has continuously been monitored for identification reasons in order to study the behavior of a karst system, as from studies formerly undertaken. Chapter 4 was focused on methods of system engineering for the analysis and identification of Fuenmayor's hydrological system, under the hypothesis of a black-box model that disregards the information on the internal structure of the aquifer. Linear and Time Invariant techniques were applied to the monitored data, and the simple correlation, spectral analysis, and parametric identification produced some interesting results. These tools have historically been successful in the study of a large number of karst springs and are a good first approach in the attempt to get a generalized model of the system. Because of the commonly accepted notion that karst is non linear and non-stationary in nature, some more precise and powerful analytical techniques are used to cover certain aspects of the analysis that the linear method paradigm leaves imprecise. Wavelet analysis, with properties such as the ability to represent simultaneously time-frequency domain, is used to identify non-stationary time-frequency relations, such as those given between input parameters and the discharge of this karst aquifer.

The goal achieved in this chapter is to characterize the behavior and response of Fuenmayor karst spring. A first analysis estimates the suitability of linear system tools, complemented with Wavelets to clearly reveal nonlinear behavior. Finally, with a Hammerstein-Wiener model, after adjustment and verification of various configurations, we arrive at a model with an efficiency value of 92 %.

6.3.2 Analysis and Identification of Collins Glacier

The aim of Chap. 5 was analyzing and modelling the relationship between the water discharge and the air temperature of Collins glacier, a subpolar type glacier located at King George Island, South Shetland, Antarctica. In spite of complexity, the hydrology of the glacier streams was captured by different mathematical models that emulate the melting-discharge dynamics. The wavelet coherence analysis has allowed for the delimitation of the transition periods at the beginning and end of any summer season. Each annual time series must be methodically prepared in order to obtain the SEC data, free of outburst effects and other anomalies. Plus, the parametric identification used to obtain yearly black box models of the glacier has revealed that one of main non linearities detected in the glacier behavior is the phase change of water. The glacier exhibits two different forms of behaviors (overdamped and overshoot) depending on the average temperature. Finally, a coalescent model, taking into account all the years covered by the study, has been identified for this glacier, in an attempt to capture its general behavior. The parametric identification, with a model efficiency about 70 %, has proved to be a powerful analysis technique, although its predictive power is limited.

6.4 Findings

The main findings in this thesis are:

- Effective methods have been tested to unveil features in the signals of an aquifer that otherwise, by classical methods, would have remained unnoticed.
- The model is built for each season of the period considered in the same aquifer site, but for the same precipitation in different years, the aquifer presents a different flow regime.
- Models in Fuenmayor spring forecast the discharge effects for short and long prediction horizons whenever the precipitation data are known.
- In some cases the complexity of nonlinear models and their computational implementation can compensate for the improvements over linear models, but in others the estimation of linear models is sufficient.
- There are procedures to quantify the interaction between climate and hydrological cycle through correlation analysis.
- From system identification, the LTI model can gather the dynamic behavior of a subpolar glacier.
- SEC represents the active period of a glacier, and its bindings have been estimated by computing the wavelet coherence.
- Among a collection of tested structures, the Hammerstein-Wiener model is a suitable nonlinear structure for the annual cycle evolution of a glacier.
- According to the SEC average temperature, the glacier dynamics exhibits two different forms of behaviors: overdamped and overshoot.
- It is possible to achieve a generalization of behavior for a particular glacier considering the merging of all annual cycle models in a *glacier coalescent model* that represent the abstraction of the glacier behavior.
- Because glaciers in this area are close to the melting point, they can play the role of change climate sensor.
- There are similarities between different systems, even more so if they belong to the same type , but that is not enough to simplify the generalization of the identification processes, so models have to be tailored or configured to each case.
- A progressive and integrated identification methodology, that is adaptively applied from linear models until nonlinear and block models, is a well undergone proposal to carry out the study of any natural system dynamics.

6.5 Contributions

The present thesis tries to provide a data-driven model framework for hydrological systems in the field of karstology and glaciology. The framework is significant because it incorporates an evaluation from the reliability of the classical system identification methods that confer legitimacy to the results. Some additional contributions have been explained and tested, e.g, the new adaptation of wavelet techniques in cases of detections of singularities in time series or irregular trans-seasonal periods.

While the statistical properties of different time-domain objective functions are well understood and applied, the implementation of wavelet spectral analysis in hydrology is still relatively recent. For this reason the analysis and model estimation in the wavelet techniques is an improvement in this field and an innovative contribution.

The findings of this thesis have meant a few scientific contributions produced over several years of research development. The most important results have been published in conference proceedings or scientific journals.

The inventive fact of this project is in finding new approaches to nonlinear models adapted to the karst and glacier system. Wavelet analysis is a fast and effective means to characterize the discharge, the air temperature, and precipitation regimes as well as their interrelationships and inter-annual variability.

From the results some immediate utilities may arise, such as an improvement in the management of karst springs or a mitigation procedure for climate change impacts. Analysis of hydro-meteorological relation in glaciers helps in the understanding of climate change impacts, available water resources in a region, melting and flow generation processes, sediment and transport processes, and natural hazards (floods, outbursts, land or ice slides, and avalanches).

Probably a better position in the melting glaciers and discharge modeling is increasingly addressed by some of the runoff models based on data derived from spectral analysis, due to the high performance in computing when implementing models with good efficiency and the fact of requiring only statistic data.

This work contributes markedly with progresses in system identification techniques that can help in the understanding complex systems, such as those related to karst and glacier hydrology. Progressive and integrated identification methodology is the proper guidance to address the study of a system with a strong stochastic component and an intrinsic unawareness of internal logic.

Moreover, data-driven models can meet other important objectives in nature environmental field, such as recovering lost data, data mining simulation, improving the reliability of predictions, and supporting references to contrast results with other estimated models.

6.6 Discussion of Results

Many data-driven models in hydrology have been criticized for their black-box characteristics, which impede adequate understanding of their detailed behavior of the system and require high heuristic support so as to be correctly interpreted.

Because a model is a simplified concept of a real system, there is not a unique model for a given groundwater system. Different assumptions compel the selection of different model structures or at least modify the parameters of a designed pattern. Hence, clear criteria must be established to determine which method is best suited for a given aquifer.

The selection of the appropriate conceptual model for the given problems was not set in the initial stage of the investigations. Instead, modeling has been performed as a continuous activity in which several alternative structures have been tested and reexamined. Nevertheless, these processes involve tedious activities in adding, deleting and modifying parameters, after checking the efficiency of each model. This leads to continuously rethinking and involves very long tasks with a remarkable computational time.

The decision for the selection of a preferred solution for data modeling typically is based on an efficiency criterion for a set of test data. Nevertheless, each criterion may place different emphasis on different types of simulated and observed behaviors. On the other hand, the simplification in testing new structures by reducing the parameters can disregard variables that are assumed to have little effect on the process.

For ease of kindness or acceptance of the method, in addition to proposals accompanied by demonstrations or tests, results from the models have been verified with those monitored previously. Nevertheless as the historical time series is not long enough—five years for the case of Fuenmayor spring and nine years in the case of the Antarctic glacier—the verification of model efficiency is subject to the availability of data.

When faced with a problem of nonlinear systems engineering, the first approach is usually to linearize the problem, or put another way, to try to avoid nonlinear aspects of the problem. This may be a good way to proceed in some cases if the fit tolerance in the validation process is agreed as a solution. However, if the solution cannot be accepted in any way, tackling the difficult problem of nonlinearity is definitely inevitable. The delicate decision is to set the level of tolerance to accept a linear identification. In this thesis multiple tests on a variety of structures have been carried out by procedures of iteration and configuration of parameters of candidate structures, though the main drawback is the lack of a comparative study to demonstrate the goodness or reliability degree of one method versus another.

There are attempts to predict long-term hydrologic change where the basin geology, the climate change and the hydrologic response are evolving in a coupling process of feeding back on each another, with the historical data as a guide to the future response. Nevertheless, the current hydrological behavior (including both bedrock structure, weather and hydrologic response) cannot readily be extrapolated to predict future behavior. This implies serious caveats in designing a model within a strict stringency, a responsibility in the verification criteria and interpretation of the foretold results. The structure and physical laws that govern the system cannot be previously known, because it is not possible to visit the Fuenmayor karst system, and in the event that it were, the physics of the system might not be accurately determined. Analogously, in the Collins glacier not all factor in the evolution of subglacial processes can be accurately determined, without descending moulins, visiting glacial caves, or going down through crevasses. However, prior knowledge of some aspects of the conceptual model might be closer to a gray box model that somehow would guide the identification and facilitate the interpretation of results obtained in the data-driven model. Therefore, systemic analysis would be a tool to indirectly help describe the physical structure of the system.

The development and application of a mathematical model should be an exercise in thinking about how a groundwater system works. Models must be regarded as a tool to aid decision-making, but decisions should not be made exclusively from results generated by the model. If the basic principles of groundwater flow or ice melting (physics laws) are lost sight of, there is a serious danger of gross misinterpretation of model output. This is more likely to occur when models are packaged as a plug-and-play automated system. In the application of all models, and most especially of groundwater karst and glacier systems, a high degree of scientific judgement moderated by a wide experience of field observations is desirable in order to produce reasonable interpretations.

6.7 Perspectives

6.7.1 Suggestion for Further Research Lines

Given the experience achieved, some suggestions are included for future research. They are issues that have arisen over the doctorate period , and even they had begun to be studied as an alternative or complementary work to enrich the core subject of this thesis. Nevertheless, in compliance with the quality of the thesis in terms of dedication, scope and dimensions, these issues should be treated in further subsequent works.

- Although already treated superficially, it would be interesting to consider the influence of other factors that could involve the discharge of Fuenmayor spring, such as humidity, atmospheric pressure, and air temperature in the inputs, and water conductivity and water temperature in the output, following the methodology implemented in this study of correlation between precipitation and aquifer discharge.
- Similarly, the other time series data collected in Collins glacier, such as the humidity, atmospheric pressure, and the radiation, could be studied in order to create a discharge model based on different glacier parameters or even a model with an interaction of multiple input signals.
- In the light of the spots appearing on the spectral analysis, an interesting research project would be to inquire into the possible correlation between SEC's glacier discharge and moon phases, or the relation with SAM effects.
- It would be interesting to implement the obtained models to test other similar karst systems, or in the case of a glacier, to verify the characterization for the same type of subpolar glaciers.
- Moreover, using different model structures for the components of Hammerstein-Wiener could improve the results, or proposing a different approach through Artificial Neural Networks.
- The integration of wavelet analysis into soft computing, with fuzzy logic and multiresolution analysis can be used to infer the rules in the frequency domain. E.g., this could be an alternative to set the SEC bounds of a glacier.

6.7.2 *Final Remarks*

A large volume of literature already exists on nonlinear models and there is still a huge amount of work to be done in spite of that. Rugh [1] opined referring to the engineering of nonlinear systems: *Those engineers who achieved significant progress are often seen as senseless or worse. The engineering of nonlinear systems is considered not only as an adventure hard and confusing, in addition, it is commonly seen as dangerous for those who think about it for a long time.* After presenting a list of techniques applicable to non-linearity as the theory of differential equations, functional analysis and operator theory, Rugh [1] went on to affirm the necessity of occupying an intermediate position between generality and applicability:

> *The nonlinearity theory can be of great importance for serving as a starting point for both the more esoteric mathematical studies and for the development of engineering techniques. It can actually serve as a* **bridge** *or communication link between these two activities.* Rugh [1] **This thesis attempts to be a little but solid stone in this bridge stretched on systems engineering and natural systems.**

Reference

1. Rugh WJ (2002) Nonlinear system theory. The Johns Hopkins University Press, Baltimore (web version 2002), (ISBN O-8018-2549-0)

Appendix A
Complementary Illustrations About Results

This appendix contains samples of analysis results in the form of illustrations, which have not included in the text of the chapters due to their complementary character or oversizing (Figs. A.1–A.8).

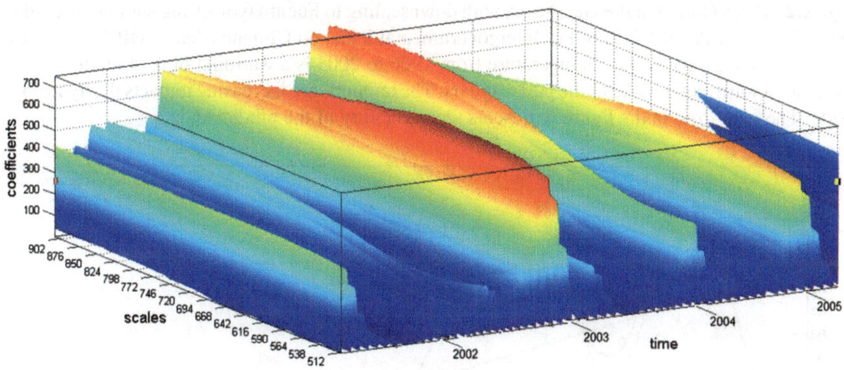

Fig. A.1 3D representation of Continuous Wavelet Transform of discharge signal in Fuenmayor aquifer

D. Chinarro, *System Engineering Applied to Fuenmayor Karst Aquifer (San Julián de Banzo, Huesca) and Collins Glacier (King George Island, Antarctica)*, Springer Theses, DOI 10.1007/978-3-319-08858-7

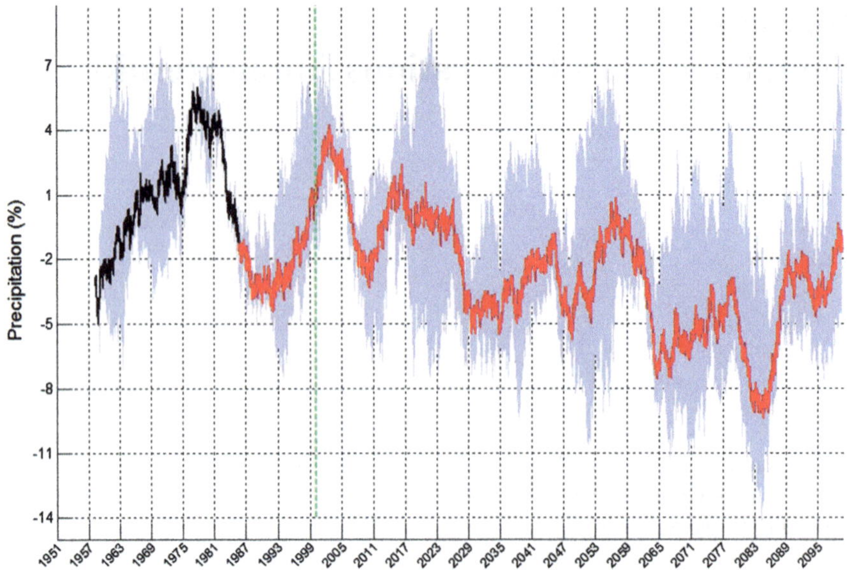

Fig. A.2 Precipitation in the Guara area with downscaling to Fuenmayor spring station. According to the Technical Paper VI (2008) of Intergovernmental Panel on Climate Change (IPCC), the best-estimate in the global surface temperature from 1906 to 2005 is a warming of 0.74 grades C, with more rapid warming trend over the past 50 years. The temperature increasing impacts the hydrologic cycle. The *gray area* in the figure represents the variability of the predicted data

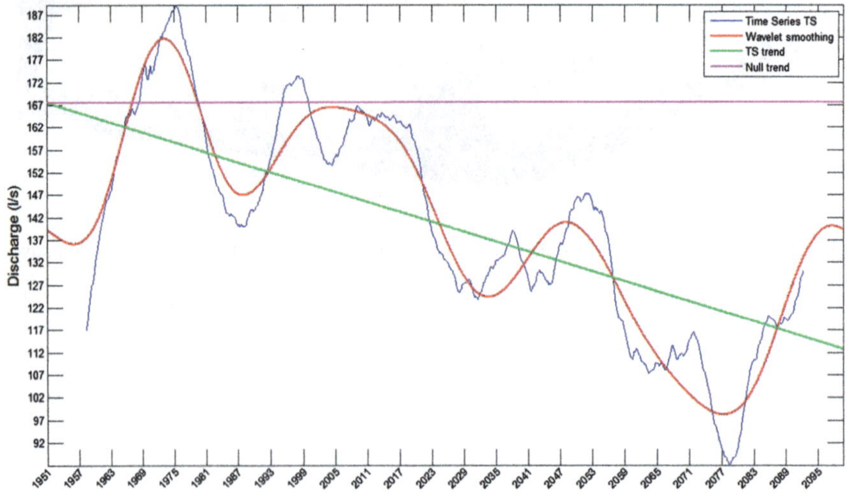

Fig. A.3 Forecasting of the discharge in Fuenmayor spring. Estimation of future trend applying Global Climate Models, downscaling methods for Aragon [1] and Hammerstein-Weiner model of this thesis (Sect. 4.6)

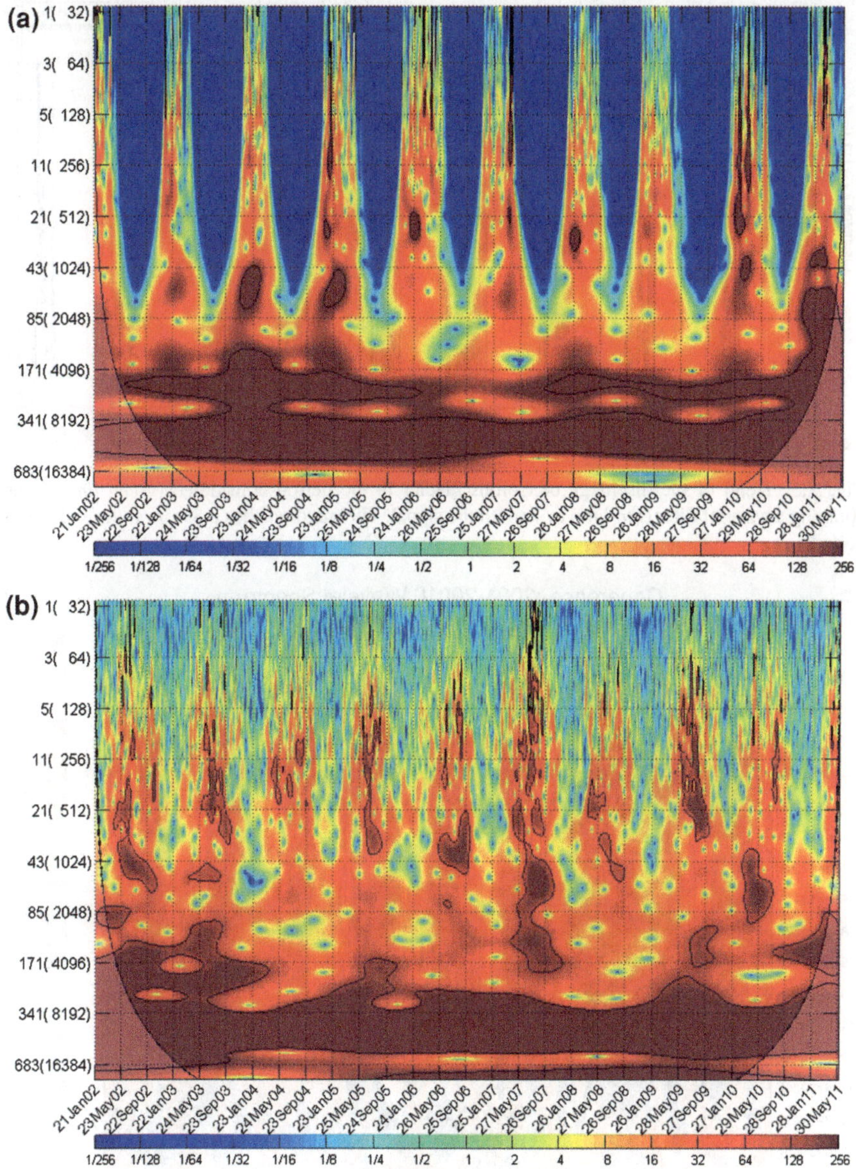

Fig. A.4 The discharge (**a**) and the air temperature (**b**) wavelet spectrum at the glacier Collins. The vertical axis shows a logarithmic scale from 2^5 up to 2^{14} h. Numbers outside the parentheses indicate the equivalent in days. The abscissa axis represents the timeline labeled quarterly, although the calculation of spectral power is held every hour. Values of the spectral power are depicted by a distribution of color hue depending on the moment in time series and the frequency (or scale). The cone of influence indicates the useful area of the spectrum without influence of the edge effect

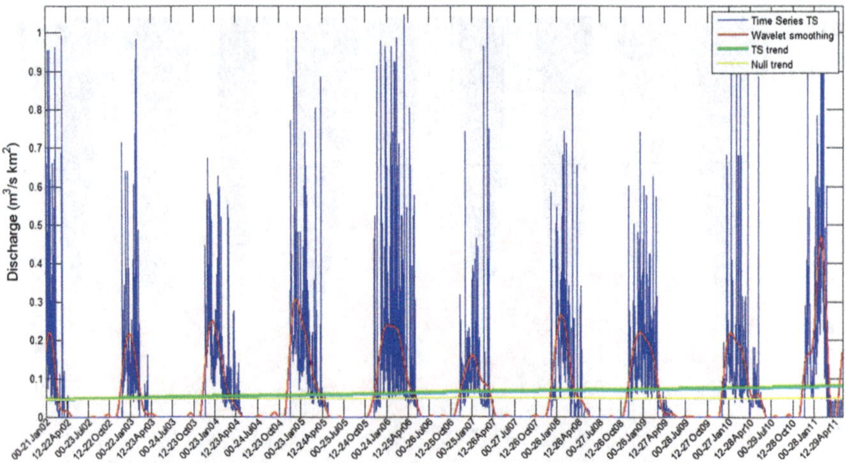

Fig. A.5 Trend of the discharge of Collins glacier in the period considered (*green line*), after applying a wavelet filter (*red line*) to smooth the signal. The trend reaches $0.8\,m^3/s$ at the end of the period

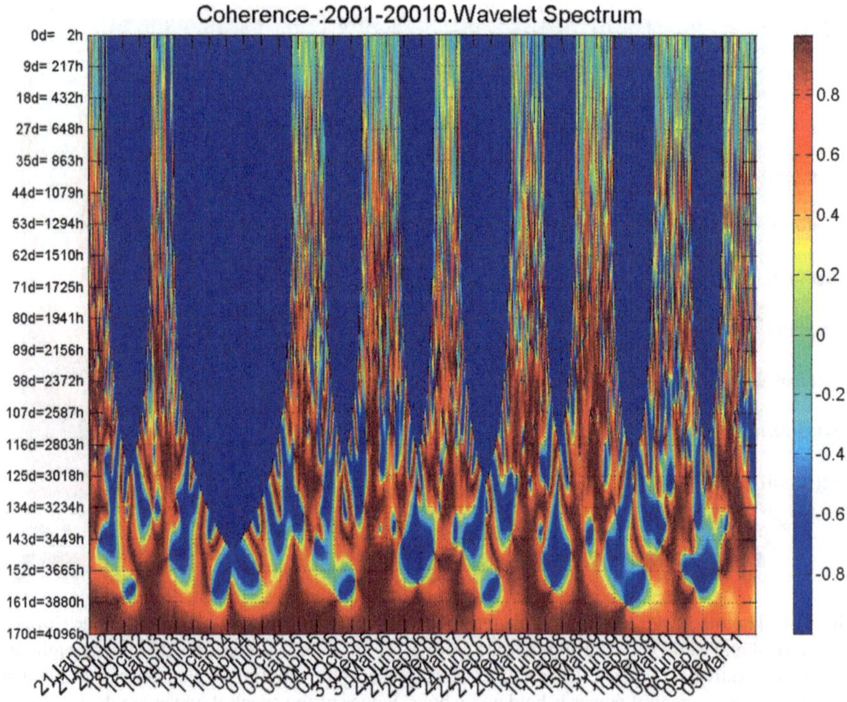

Fig. A.6 Wavelet spectrum coherence between air temperature and discharge of Collins preglacial stream

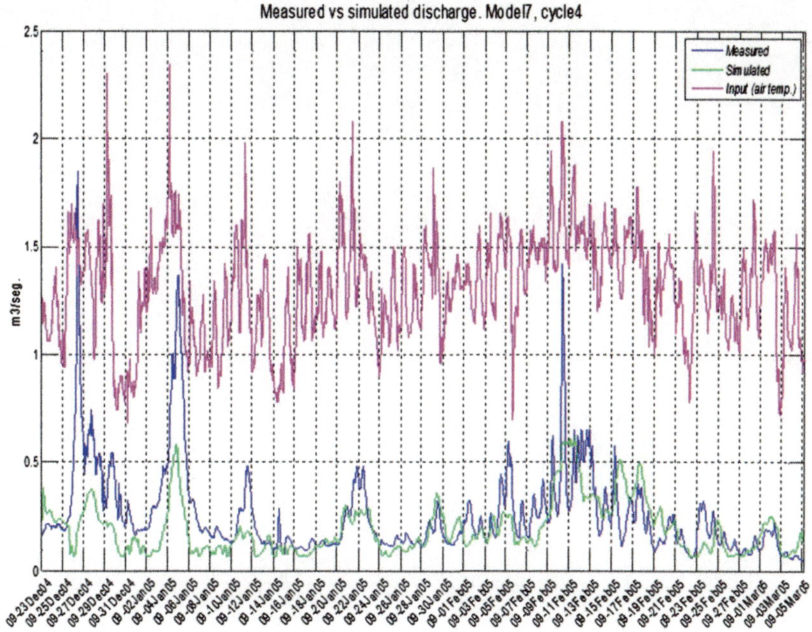

Fig. A.7 Collins glacier models. Verification of Model 8 (Hammerstein-Wiener) with messured data during cycle 2 (2002–2003 season)

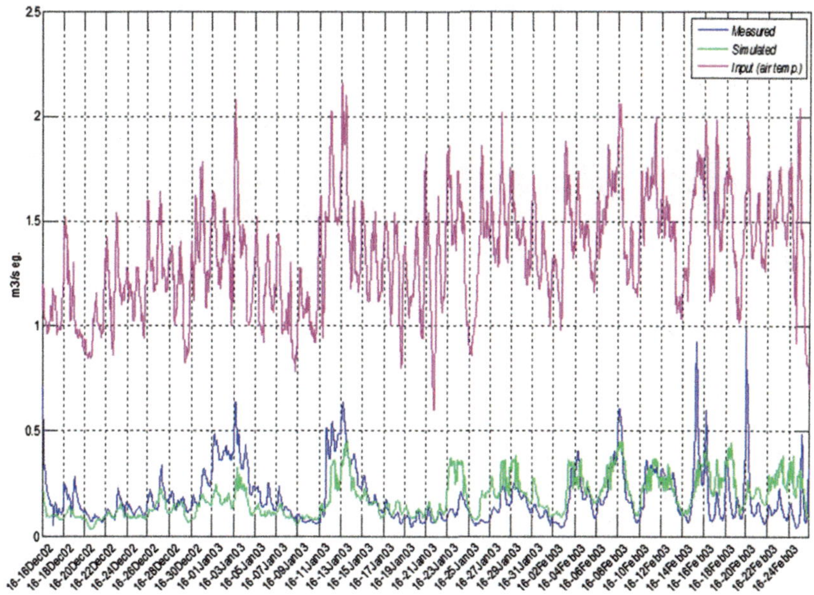

Fig. A.8 Collins glacier models. Verification of Model 7 (Hammerstein-Wiener) with messured data during cycle 4 (2004–2005 season)

Appendix B
Glossary

Some terms have been included in this glossary by alphabetical order, defined according to the meaning used in the thesis context, and based on these references:

Cave and karst terminology by J. N. Jennings, Australian Speleological Federation Inc. 1997; *A Lexicon of Cave and Karst Terminology with Special Reference to Environmental Karst Hydrology* by United States Office Environmental Protection Agency, EPA/600/R-02/003, February 2002, www.epa.gov/ncea; *Glossary of Glacier Terminology* by U.S. Geological Survey, http://pubs.usgs.gov/of/2004/1216/text.html, 2013; *Glossary of Glacier Terms* by Ellin Beltz, Illinois University Earth Science, 2006; *Dictionary of Scientific and Technical Terms*, by Sybil P. Parker, McGraw-Hill, 2002.

Aggressiveness Features of water capable of dissolving more limestone or speleothems by the action of other chemical processes (hydration, ionic substitution and oxidation-reduction), and physical phenomena (mass transfer and diffusion).

Autocorrelation A technique used to detect cyclic activity in a complex signal.

Avalanche An avalanche (or snowslide) is a rapid flow of ice or snow moving surprisingly down a slope, because internal melting and mechanical failure of the snowpack.

Bédière (gallicism) is a stream that runs through the surface of a glacier and is fed by snowmelt.

Black box is a device, system or object which can be viewed in terms of its input, output and transfer characteristics without any knowledge of its internal workings.

Complex system A system that has multiple parts that interact to produce results that cannot be explained by simply specifying the roles of the various parts.

Convolution Convolution is the operation on two functions f and g, that can be calculated by integral: $f(t) * g(t) = \int_{R^n} f(t)g(t-\tau)d\tau = g(t) * f(t)$ and describes the weighted average of the function f(τ) at the moment t where the weighting is given by g(-τ) simply shifted by amount t.

Crevasse Elongated open cracks in glacial ice, usually nearly vertical, and subject to change at any moment.

D. Chinarro, *System Engineering Applied to Fuenmayor Karst Aquifer (San Julián de Banzo, Huesca) and Collins Glacier (King George Island, Antarctica)*, Springer Theses, DOI 10.1007/978-3-319-08858-7

Cryosphere The perennially frozen regions of the planet, including land-ice, sea-ice, permanent snow cover and permafrost.

Dirac delta is a mathematical construct-not strictly a function- introduced by Paul Dirac—that has the value zero everywhere except at x = 0 where its value is infinite.

Discharge The total volume of aquifer water from an aquifer or melting water from a glacier ice passing through a specified cross section of the drenage stream during a particular unit of time.

Dome The high, central stable portion of an ice sheet with low rates of accumulation and slow ice movement.

Dynamic system A set of equations specifying how certain variables change over time. The equations specify how to determine (compute) the new values as a function of their current values and control parameters. The functions, when explicit, are either difference equations or differential equations.

Equilibrium line The boundary between the accumulation area and the ablation area.

Evaporation Endothermic physical process in which liquid water changes into water vapor (100 C at 1 atm). Absorbs 540 cals/g latent heat.

Fissure A deep, long and narrow opening such as a crevasse in a glacier.

Glaciation Period of time during an ice age when glaciers advance because of colder temperatures. (2) Involving glaciers and moving ice. Usually pertaining to processes associated with glaciers.

Ice cap Smaller ice sheets, with radial flow, which cap many islands in the Arctic and Antarctic regions.

Ice sheet A mass of land ice of continental size, thick enough to cover the underlying bedrock ground, and shape is mainly determined by the dynamics of its flow.

Karren Minor forms of karst due to solution of rock on the surface or underground. Lapies is the same thing, the term originates from the French. In English both terms are used equally and synonymously.

Limestone A sedimentary rock consisting mainly of calcium carbonate, $CaCO_3$, derived from the accumulated deposition (and fossilization) of the calcareous remains of marine or freshwater organisms.

Linear system When the output depends on the input by a linear equation of the form y = mu + b, in which y varies "linearly" with input u.

Moulin A narrow, tubular chute, hole or crevasse through which water enters a glacier from the surface. The term is derived from the French word for mill.

Noise In experiments, noise usually denotes the small random background disturbance of either mechanical, thermal, or electrical origin

Nonlinear dynamical systems A system whose behavioral output can suffer multiplicative or exponential effects respect to input, such that small perturbations can carry the capability to trigger substantial changes in the output, or conversely, large perturbations can alter the system's trajectory only slightly.

Nunatak An Eskimo word for mountains or lands which protrude through the ice.

Outburst A sudden release of melt water from a glacier or glacier-dammed lake sometimes as a result of he instability of the snowpack during transitional seasons.

Polje A large closed depression draining underground, with a flat floor across which there may be an intermittent or perennial stream and which may be liable to flood and become a lake. The floor makes a sharp break with parts of surrounding slopes.

Phreatic zone Zone where voids in the rock are completely filled with water.

Proglacial of streams, deposits, and other features, being immediately in front of or just beyond the outer limits of a glacier or ice sheet, and formed by or derived from glacier ice.

Proglacial lake A lake formed either by the damming action of a moraine or ice dam during the retreat of a melting glacier or by meltwater trapped among ice sheets.

Signal Function of time representing some variable that contains information about the behavior of a natural system.

Speleothem are cave features created after the underground chamber has been formed. They are a result of slow-moving water, usually containing calcium carbonate, which has been dissolved from the limestone where the cave was formed.

Subglacial The area below the glacier with features include deformed sediments, ice caves, and eskers.

Systems Engineering The orderly process of bringing a system into being using a systems approach.

Systems Engineering Process is an iterative problem solving process based on the fundamental cycle of analyse-synthesise-evaluate.

Systems Engineering Tools are those tools that support systems engineering processes.

Tozal Term used in the Aragon region, that means a high and lofty place (from Arabic).

Trop plein (from French 'too full, overflow') is a karst upwelling groundwater unusual in nature—although many of them go unnoticed when being covered by vegetation, stony, or other sheets higher water—produced when the aquifer is too full and overflows by discharging on higher levels, through a rocky passage already formed.

Vadose zone Zone where voids in the rock are partly filled with air and through which water descends under gravity.

Weathering The process of physical and chemical decomposition which changes earth and rock materials in color, texture, composition, firmness or form. These changes are accomplished by the effects of energy and exposure to water, other fluids and the atmosphere.

Reference

1. Ribalaygua J, Pino M, Pórtoles J, Roldßn E, Gaitßn E, Chinarro D, Torres L (2013) Climate change scenarios for temperature and precipitation in Aragón (Spain). Sci Total Environ, Special Issue:1–13

CPSIA information can be obtained at www.ICGtesting.com
Printed in the USA
LVOW02*0319280714

396230LV00002BB/49/P